THE LASER MARKETPLACE IN 1988

SPIE Volume 950

Contents

PREFACE

Each year *Laser Focus* magazine and its affiliated business newsletter, *Laser Report,* organize a seminar to examine the worldwide market for lasers. The seminar is coordinated by *Laser Focus* Senior Contributing Editor for Northern California, Dr. Gary T. Forrest. We invite your comments and questions, which may be directed to

> Dr. Gary T. Forrest
> FYI Reports
> 491 Seaport Court #101
> Redwood City, CA 94063
> 415-361-8006

This year's seminar was held on 13 January 1988 at the Los Angeles Hilton Hotel in conjunction with the SPIE O-E/LASE '88 meeting. Attendees received a full set of speakers notes for the day-long seminar. This volume is a direct transcription of the proceedings, including selected viewgraphs used by the speakers.

Twelve speakers covered both general and application-specific markets in half-hour presentations. The speakers and their presentation titles are shown on the contents page.

The *Laser Focus* staff serves as a focal point in the laser industry for both technical and business news. In addition to *Laser Focus* magazine and the more business-oriented *Laser Report* newsletter, the staff publishes *Japan Laser Report, Medical Laser Industry Report, Industrial Laser Review, Laser Focus Buyer's Guide, Industrial Laser Annual Handbook, Medical Laser Buyer's Guide, Journal of Current Laser Abstracts,* and a wide variety of review supplements that cover laser, optics, fiber optics, and imaging technologies and marketplaces. For further information contact

> Dr. Morris Levitt
> *Laser Focus*
> 119 Russell Street
> Littleton, MA 01460
> 617-486-9501

We welcome your comments and your personal attendance at the 1989 seminar. About half of the topics for each year include core topics such as emerging laser technologies, market overview, and scientific, industrial, and medical market reviews. In addition, about half of the topics are selected for their timeliness in a particular year. We welcome your input on such topics for future presentations.

Morris Levitt
Gary T. Forrest

Books are to be returned on or before
the last date below.

LIBREX–

The   1988

SPIE—The International Society for Optical Engineering

Held in conjunction with SPIE's O-E/LASE '88
13 January 1988
Los Angeles, California

Published by
SPIE—The International Society for Optical Engineering
P.O. Box 10, Bellingham, Washington 98227-0010 USA
Telephone 206/676-3290 (Pacific Time) • Telex 46-7053

SPIE (The Society of Photo-Optical Instrumentation Engineers) is a nonprofit society dedicated to advancing engineering
and scientific applications of optical, electro-optical, and optoelectronic instrumentation, systems, and technology.

Please use the following format to cite material from this book:
   Author(s), ''Title of Paper,'' *The Laser Marketplace in 1988,* Morris Levitt, Gary T. Forrest, Editors, Proc. SPIE 950, page numbers (1988).

Library of Congress Catalog Card No. 88-61004
ISBN 0-89252-985-7

SPIE Volume 950

THE LASER MARKETPLACE
IN 1988

*A seminar examining recent trends and directions
in the worldwide market for lasers*

Editors

Morris Levitt
Laser Focus

Gary T. Forrest
FYI Reports

LASER MARKET OVERVIEW
1987 RESULTS AND 1988 FORECASTS

Morris Levitt
Editor-In-Chief, Laser Focus Magazine

This morning I will review recent Sales and Market developments for each principal laser type and application market, and then discuss the projections in most of the major markets looking ahead through the end of 1988. In each of these major areas, there will be more elaborate data and analysis presented by the speakers who are specialists and industry leaders in each of these fields.

My presentation is primarily based on the reports that you have in hand, or are available to you at the table outside. That is to say the annual economic review and outlook that appears in the January issue of Laser Report and in more summary form in Laser Focus that was prepared by our business editor, David Kales, who will be speaking later on a related subject. So, much of what I have to say is based on the methodology that we have traditionally used through the years in the annual Laser Focus and Laser Report reviews, which is to go to the primary manufacturers and suppliers and interrogate them as to their own sales and their own projections of the worldwide market in each of the principal areas of laser devices and markets; to compare that against other extant reports that are available, and try to achieve a maximum of self-consistency within all the pieces of data that we have available to us. Obviously the optimal way to do this would be to go to every consumer of laser devices, but unfortunately, by the time such data were available, the market would have probably passed by by about eight years.

If we begin with the broadest bottom line results for the year, in 1987 worldwide sales of commercial lasers, and by that we mean the non-Communist world, we estimate grew about 9% to a level of $570,000,000 in the past year. Our forecast calls for these sales to grow at about the same rate in 1988, (by about 8%), which would get us up for the first time over the $600,000,000 level, to about $613,000,000. And, of course, there is a caveat that we have to use in all of these projections since we are probably in one of the most volatile, or uncertain, economic periods we have been in for some time, both from a purely financial standpoint and in terms of national and international political decisions and how they might impact economies and markets. There will always be at least a reasonable chance of recessionary trends in the year or two ahead, and that could throw into a cocked hat all of these projections that we have done on a more microscopic level looking at the laser industry per se.

What these results tell us more broadly, as indicated in one of the up-front punch lines in Laser Report, is that it would appear that the days of double-digit growth for the laser industry may have forever passed from the scene. It is also important to note that the results that you see here do not reflect any monolithic trend. These results are really the average over a number of quite disparate markets. As we will see, we are looking at some markets that are severely down, some which have stayed relatively flat in the last year, several which have grown at historical rates, and even a few, thank God, that are indicating explosive growth and continued potential for that type of growth.

So what can we make of the implications of these kinds of numbers, how would we characterize, or how would I characterize some current health index for our industry? I think in order to answer that question, we have to note that in addition to being also widely different mix of markets, the animal we call the laser industry is also a widely disparate group of players. It continues to consist of very broad-based companies such as Spectra Physics and Coherent which manufacture across the board types of laser devices. But even within that category we now have companies that are independently owned financially, as well as companies that are now subsidiaries of Fortune 500 type companies, and therefore some are public and some are private. We have a number of laser companies that I think are more appropriately described as niche companies, whether with respect to the products they manufacture or with respect to their business strategies in terms of seeking to penetrate just a few select markets. And again, even within that stratum of companies, we have those that remain under independent management whether public or private, and those that are subsidiaries of much larger corporate entities with more far ranging interests.

We also have a group of companies that are not simply laser suppliers but from the standpoint of also going after value added, or because they have historical roots in particular industries, may be system manufacturers. And within that group of companies there are some who make their own lasers that go into those systems, others who are important OEM consumers of lasers, and again within those levels of differentiation, there may be companies that are privately held or independently managed, and those which are subsidiaries of the types of companies we read about in Business Week.

So, essentially what we are faced with in looking at what lies behind that data is a mosaic of companies with a very complex dynamics. From time to time there are analysts who project "what will the industry look like in ten years" and try to project a much more simplified version in which all companies will be integrated vertically or as part of Fortune 500 companies, or divided along lines of small niche companies. I personally don't that such simplification will ever come to pass, although there are trends in that direction. Especially in the U.S., I personally find that the laser industry tends to reflect the much more complex corporate and market dynamics that we have in the U.S. say, compared to Europe and Japan.

Putting that altogether, my own view of the current state of the laser industry is that it is in very healthy shape, despite this apparent slow down in revenues. I say that from the standpoint of perhaps three principal factors:

1) The first is that, as we will hear, for example, in Lewis Holmes' talk, I think the industry has the most solid technological base it has ever had, and there continue to be both developments of new types of devices and improvements in performance and reliability that give a very solid technical underpinning to the industry.

2) Secondly, I think we are operating from the standpoint of much better management in the industry, in which internal to companies there are far fewer delusions about how easy or difficult it may be to perform effectively in the marketplace, and also related to that, companies are making far fewer false claims to the market which would tend to undermine the credibility of our industry in the long-term.

3) And, finally, I think when you put that altogether, as we will see in some of the market segments that we examine, there is a much better fit between the technologies we are developing and the markets we serve. There is a much greater response to the needs of those markets and the technologies we are developing.

MARKET HIGHLIGHTS

This is somewhat arbitrary, but to look at some of the broad highlights in terms of key developments in the past year, first we have to note that on the surface the material processing market appears to be flat to down. This looks like an extremely difficult environment in which to operate. This is primarily due to the fact that there has been a sustained softening of the capital goods sector worldwide, and there has been a severe turndown in the micro-electronics and semi-conductor area. But there may be other factors such as how good our data or our analysis are. I'm not going to say too much more on that because my colleague Dave Belforte, who is very close to this market, I think may have a unique view of where this marketplace is going.

The medical area, by contrast, was up a strong 17% in the last year, although that was made up of several contradictory factors. The therapeutic area was strongly up, but within that the ophthalmic marketplace is saturated and relative to the two or three dozen players that we had, particularly in the YAG area several years ago, we are down to maybe a still viable half dozen. But at the same time, there is very strong growth in the medical/surgical area, both in terms of compact CO_2 units and YAG units for endoscopy. And one of the brightest regions in terms of percentage of growth that we are looking at this year and into the future has been the growth of the diagnostic area.

More broadly, the industry witnessed a number of important consolidations and realignments this year. For example, the acquisition of Rofin-Sinar, the largest European supplier by Siemens and the stunning acquisition of Spectra Physics by Ciba-Geigy. There was a major session held yesterday by the Laser Association of America on the impact of the Gould patents held by the Patlex Corporation. There were definitive legal decisions made this year after perhaps a decade of inconclusive legalities, which have now put the entire industry in the position of having to pay very significant royalties to Patlex Corporation, which may very well, and already has in fact, affected bottom line performance and may impact the financial results in the industry for years to come.

In October the laser industry shared, with the rest of American industry, in being severely impacted at least in terms of financial evaluation, by the stock market collapse. I will show you some numbers on that in a minute.

All in all, however, I think it is safe to say that the laser industry weathered this set of economic downturns as it has in the past, because of its broad diversification and its incessant ability to seek out new market niches. And we have also witnessed in the past year, a number of areas that had extraordinarily strong growth, among them diagnostic medicine, the broad R&D area, barcode scanning, and optical memories -- all of these underscoring the industry's resiliency.

From a technology standpoint, which Lewis Holmes will address, there were very significant advances in device technology in such areas as diode, solid-state, diode-pumped solid-state, excimer, ion, and even copper vapor and He-Ne devices.

From the purely financial or corporate standpoint, however, unfortunately I have to report that we continued to see profits proving to be extremely elusive (Fig. 1). If we look at figures available for some of the principal public companies in the industry, Coherent, for example, which closed the books on its fiscal year in September, even though it had record sales growth up to $159,000,000, still managed to lose over a million dollars on those record sales. And there were factors at play there such as the softness in the industrial area. Lumonics hit the mid-year point in its fiscal year in September. It had about the same revenues as last year - $50,000,000, but managed to lose a million instead of the profitability of half a million the previous year. Again one of the key factors may be the problems in the industrial area. Quantronix at mid-year was down in both revenue and profitability. This is an example of a company that may have taken a double hit in terms of problems in one of its key markets, the semi-

Figure 1
Laser Company Financial Performance, 1987

Company	Sales	Earnings
Coherent (Sept.)	$159M (+15%)	-$1.1M
Lumonics (9 Mo./Sept)	$50M (n.c.)	-$1M ($456,000 in '86)
Quantronix (6 Mo./Sept.)	$9M (-18%)	-$1.3M (Loss due to patent payment to Patlex.)
Laser Photonics (9 Mo./Sept)	$3.7M ($3.6M in '86)	-$48,600 (-$1.7M)
Laser Industries (6 Mo./Sept.)	$18.5M (+25%)	$2M (n.c.)
Patlex (9 Mo./Sept)	$2.9M (Project $17M for '88)	-815,400

conductor, and then was right there when the Patlex train came down the rails. Laser Photonics hit the three-quarter mark in September at about the same revenues. It improved its profitability to almost break-even from major losses of the previous year. Laser Industries, an Israeli company which is traditionally one of the most profitable in the industry, had very aggressive 25% growth in revenues but no change in its profit at about two million. (I'll take it). Patlex, which is probably one of the more interesting financial plays in the industry, was at the nine-month level in September. Revenues were only about three million, but they are projecting as much as seventeen million dollars in revenues next year, even though they lost over eight hundred thousand dollars this year. If they achieve those revenues they are projecting, they could become the company with the highest earnings in our industry.

These results were reflected in the pre- and post-stock crash price levels. These appeared in Laser Report (Fig. 2). Basically there is a mix of conditions here. Some companies who were performing well managed to stay even on the stock index for the year because after taking their hit in the crash, they still had enough strength to share in the recovery. So companies like Coherent are about the same as they were. Patlex is about the same as it was, but other companies like Quantronix and Laser Photonics are still down significantly. And I would advise those of you who like the crap-shoot called the stock market, that some of those may be interesting buys at this point.

If you look at how these laser stocks compare to the stock market in general, in the middle of that chart of mutual funds performance, there is an obscure fund called Growth Laser Fund (Fig. 3). I guess this summarizes some of the financial problems that we are discussing. The laser fund was about the fifth or sixth worst performer, both in the fourth quarter and for the year, which means that the broad stock index for laser companies was one of the few that did not participate in the bull market and it was also one of the few that didn't partici-pate in the recovery. That reflects that we are a highly volatile and not yet highly profitable financial entity as an industry.

THE FORECAST FOR 1988

Looking ahead to 1988, it would appear that this softness in the industrial market will continue, at least based on the kind of data that we are gathering. Again, I will indicate that before you draw too many strong conclusions, you should wait to hear what Dave Belforte has to say. This has impacted both CO_2 and YAG lasers, which reflects the fact that we are looking at softness in both the capital goods metal working sector as well as the micro-electronics and semi-conductor areas. This is an extremely tricky market, both to measure and analyze however.

We look for much more positive results in 1988 in the medical area. We expect that the therapeutic devices will slow down somewhat, but could still expect to post a 9% gain. This market is still ham-pered by the softness in the ophthalmic market, but we expect fast growth in such sectors as portable CO_2 surgical lasers, endoscopy, and one of the brightest spots and hottest areas, and a very positive area for the ion lasers in particular will be the continued strong growth of

Figure 2

Pre- & Post-Crash Laser Stock Prices

PATLEX STOCK JUMPS--MOST OTHER LASER STOCKS INCH UP

STOCK MARKET REPORT
(As of December 23, 1987)

	bid	close	change
Candela Laser	6 1/2		+1
Coherent		10 3/8 +	1/8
Control Laser		1 1/4 -	3/4
Cooper LaserSonics		9/16 -	3/16
Laser Industries	6 3/8		+ 3/8
Lasermetrics	1/2		+ 5/16
Laser Photonics	1 1/8		+ 1/8
Laser Precision		3 3/8 +	1/8
Lasertechnics	1 3/4		+ 9/16
Lumonics		5 1/8 +	1/8
Patlex		13 1/2 +	4 1/2
Quantronix		2 5/16+	9/16
Trimedyne		10	+1

STOCK MARKET REPORT (as of December 22, 1986)

	bid	asked	change		bid	asked	change
CK Coherent	11 1/2	11 5/3	+ 5/3	Laserscan Intl	--	--	--
Control Laser	4 5/8	4 7/8	- 3/8	Lasers for Medicine	1 1/4	1 3/8	- 1/8
Cooper LaserSonics	2 1/4	2 3/8	- 1/16	Lasertechnics	1 13/16	1 15/16 -	10/16
Candela Laser	4 5/8	4 7/8	- 2 1/8	Lumonics	8 3/4	--	- 1/4
JEC Lasers	3/8	--	--	Patlex	13 1/4	13 1/2	- 1 3/4
Laser Industries	11 7/8	12	- 3/8	Quantronix	6	6 1/2	- 3/4
Laser Photonics	2 7/8	3	- 4/8	Spectra-Physics	18 1/8	18 5/3	- 1 3/4
Lasermetrics	5/8	3/4	--	Trimedyne	10 1/4	10 1/2	- 1 3/8

Figure 3

Mutual Fund Performance, 1987

Best and worst performing mutual funds

Top performers - 4th quarter		Worst performers - 4th quarter	
Oppenheimer Ninety-Ten	+43.19%	44 Wall Street Equity	-49.32%
Benham Target 2015	+25.91%	Security Omni	-39.45%
Benham Target 2010	+24.15%	American Capital OTC	-39.00%
Intl Cash-Yen Csh	+19.48%	Fidelity Select Brokerage	-38.76%
Transatlantic Income	+19.32%	Fidelity Select Energy Ser	-38.32%
Intl Heritage Ovsea In	+18.71%	Div/Gro-Laser & Adv Tech	-38.31%
Benham Target 2005	+18.68%	American Capital Growth	-37.82%
Mass Final Intl Tr-Bond	+17.65%	USAA Gold	-37.37%
First Inv US Govt Plus-I	+16.98%	Steadman Oceanographic	-37.02%
Merrill Lynch Ret Global	+16.92%	Health Monitors	-36.63%

Top performers - 1987		Worst performers - 1987	
Oppenheimer Ninety-Ten	+93.56%	44 Wall Street Equity	-41.91%
DFA Japan Small Co.	+87.76%	Kaufman Fund	-37.16%
Oppenh'r Gld & Sp. Min.	+71.59%	Fidelity Select Brokerage	-37.09%
N.E. Zenith Cap. Gro.	+52.71%	44 Wall Street	-34.63%
IDS Precious Metals	+52.51%	Strategic Capital Gains	-33.03%
GT Japan Growth	+51.74%	Div/Gro-Laser & Adv. Tech.	-32.33%
DFA U.K. Sm. Co.	+51.53%	Security Omni	-27.79%
Franklin Gold Fund	+51.50%	American Capital OTC	-25.69%
Van Eck Gold/Resources	+47.23%	Fidelity Select Defense	-23.24%
Colonial Adv. Str. Gold	+46.86%	Wealth Monitors	-22.98%

Source: Lipper Analytical Services

Globe staff chart

perhaps as high as 44% for diagnostic medical equipment.

If we turn to the old reliable segment of the industry, which is R&D, we expect that this market will increase again by 13% to nearly $140,000,000. If that happens, and if the industrial sector stays soft, then for the first time or the first time in quite a while, we will have an interesting turn of the wheel in the sense that R&D will again be the largest market for lasers. Just when we thought in the mid-70's that the days of all the exotic applications were at hand and we would leave R&D behind. There are a number of technologies that are helping to drive that new growth in the R&D market, such as diode pumped, YAG lasers, and the various spectroscopic and research lasers.

On the other hand, there is always a question mark associated with this market in that it is highly dependent on various forms of government funding. While this does directly reflect SDI spending, there is a large trickle-down effect from SDI in the sense that labs like Livermore and Lincoln Labs have a lot of cash to spend and they can buy a lot of diagnostic equipment. There are other areas serving the more purely academic facilities that are funded by the NSF, and we all know that there are going to be major re-evaluations of the federal budget over the coming years.

However, this area has always been a mainstay for the laser industry because of the laser's role as a ubiquitous scientific instrument.

The 'midas touch' in our industry goes to the diode lasers. Just about every market you can look at, that diodes are associated with, are either going to hold their own or show some wild growth. For example, in the printer market there are probably two million units of diode lasers that will be consumed in 1988, compared with the already large 1.4 million units in 1987. Which means that dollar sales, even though unit prices are going down, can get as high as eighteen million dollars just in this market. This is a fascinating market in terms of its internal dynamics as well, because at the same time the diodes are moving up strongly, He-Nes have experienced a temporary growth while ion lasers are down. The decline of ion lasers will probably continue but He-Nes are going to face sharper competition in the future across the board as improved diode lasers begin to compete for all the niches here.

Barcode scanning is another area where He-Nes and diodes are mixing it up. This was one of the fastest growing areas in the past year. We expect unit sales to get up to three hundred thousand next year, a growth rate of 30%; and dollar sales soaring by 56% to eighteen million dollars. He-Nes have had the bulk of the growth here, but this has been an extremely strong growth area for diodes as well. With the development of reliable, effectively priced, visible diode lasers, this could be another area where the immediate sharp growth of He-Nes could come under attack from diodes. But there is no question this is going to be a very strong growth area as we move beyond the supermarket to the broader retail and inventory control areas, and the use of these kinds of systems and technologies by the military.

The biggest boomer of all is found in most of our homes in the area of optical memory. Phenomenal numbers of diode lasers are being consumed by the compact disc marketplace. The dollar value will probably get up to ninety-five million dollars. The number of diodes used in compact discs next year may grow to an astounding eighteen million units. It was already fourteen and one-half million units this year. The growth rate this year was 55% and this market must be maturing because the growth will only be 27%. Another important area that we will hear about, not yet at this volume, is the market for write-once optical storage, that Ed Rothchild will be commenting on.

So if we look at all of these markets together then, let me give you a kind of fast rating index. There are basically three tiers now in terms of size of application market. The largest two markets for lasers are material processing and R&D. As we have seen, the first is slowing down, the second is still experiencing solid growth. Those are both well in excess of one hundred million dollar markets. In the second tier, which is just approaching the hundred million dollar level, are two strongly growing, in one case explosively growing, markets-- the medical and the optical memory. The third tier is communications which are diode lasers used in fiberoptics. That has been a flat market as the long-haul telephonic networks have saturated and prices have come down. But it is projected that as we move into local area network applications, (we will hear about that market this afternoon), the number of devices per kilometer in those markets will sharply increase and the market for those components will probably pick up strongly somewhere in the late 80's and early 90's. Among the markets where we can identify good, strong, sustainable growth then -- R&D, medical, printing, optical memory, barcode, test and measurement. But within these we have to be careful to note that specific generic types of lasers may be strong or under attack in one market, as compared with another. So you have to really look at the entire matrix.

Let me show you a horrible chart (Fig. 4) that I would never show you if it wasn't in the notes, and it is in the Annual Laser Market Review (in Laser Report and Laser Focus) as well. If you look at the matrix, I think ultimately the best way to do what I just said, which is to assess what is going on for each laser type by looking at its respective markets side by side, or to look within each market segment, what are the dynamics for each laser -- you can only do that with a matrix of this type. This is in terms of units, some of the most striking features -- you will see that unit sales for CO_2 lasers are projected to go down in its two principal markets, both material processing and medical. Ergo, CO_2 is expected to be a laser type in decline in the next year. Solid state lasers are expected to grow due to growth in the therapeutic medical area, although they will be declining somewhat still in the industrial sector. Ion lasers are expected to grow as a result of growth in several segments -- the therapeutic medical, the diagnostic (which is very important) and R&D. Diode lasers are the column that show the numbers that are off the chart relative to everything else. If you look at the rows for printing, optical memory, communications and barcode scanners, there is a laser type that already has astounding numbers and is showing phenomenal rates of growth in each of those four market segments, even though it is competing with other laser types in some of those segments. While He-Ne is expected to decline somewhat in the printing area, it is still growing strongly in the barcode scanning area, but the net result is still somewhat down for He-Nes.

If we turn to dollar sales, and there is a more convenient summary of this data on page 8 of the Annual Economic Review of Laser Report, you can again see that the correlation between declines in markets and declines in laser device types. Let me summarize this for you in a more convenient way. If we ask 'what were the fastest growing types of lasers in 1987', the answer is that ion lasers in this past year grew by 16%, He-Ne grew by 14%, dye by 13%, diode by 11% (and this is in dollars), solid state by 7%, CO_2 by 3% and excimer and He-Cd were about flat. Looking ahead to 1988, we get quite a significant rearrangement if we ask what are expected to be the highest percentage growth gainers in dollars in 1988. We project, due to some new markets as well as sustaining in some of the older ones, that He-Cd goes from the bottom to the top, and could grow as much as 33%. For diode lasers, the

Figure 4

Unit Sales of Commercial Lasers

WORLDWIDE COMMERCIAL LASER SALES

1987–1988 (By Unit)

	CO$_2$ 1987	CO$_2$ 1988	Solid-State 1987	Solid-State 1988	Ion 1987	Ion 1988	Diode 1987	Diode 1988	He-Ne 1987	He-Ne 1988	Dye 1987	Dye 1988	Excimer 1987	Excimer 1988	He-Cd 1987	He-Cd 1988	TOTALS 1987	TOTALS 1988
Materials Processing	1400	1200	900	800	80	130	0	0	0	0	0	0	35	55	0	0	2415	2185
Therapeutic Medicine	1400	1600	1200	1400	1900	2150	1254	1464	7000	3500	150	100	15	35	0	0	12,919	10,249
Diagnostic Medicine	0	0	0	0	600	1200	0	0	11,000	12,000	70	90	0	0	70	70	11,740	13,360
R&D	480	450	775	900	1000	1200	11,614	13,704	15,000	18,000	800	900	350	340	450	500	30,469	35,994
Printing	0	0	0	0	5000	5000	1,400,000	2,000,000	45,000	40,000	0	0	0	0	800	500	1,455,800	2,045,500
Platemaking	0	0	50	50	140	140	0	0	0	0	0	0	0	0	0	0	190	190
Color Separation	0	0	0	0	5200	5000	6	0	420	500	0	0	0	0	85	50	5705	5550
Optical Memories	0	0	0	0	80	120	14,500,000	18,000,000	0	0	0	0	0	0	260	300	14,500,340	18,420,000
Communications	0	0	0	0	0	0	100,000	140,000	90	300	0	0	0	0	0	0	100,000	140,300
Barcode Scanners	0	0	0	0	0	0	100,000	150,000	130,000	150,000	0	0	0	0	0	0	230,000	300,000
Alignment & Control	0	0	0	0	0	0	12,000	14,000	8400	9700	0	0	0	0	0	50	20,400	23,750
Test & Measurement	27	30	27	25	175	275	5405	5675	34,000	38,000	10	10	0	0	280	400	39,924	44,415
Entertainment	0	0	0	0	400	475	0	0	2600	2800	25	25	0	0	0	0	3025	3300
TOTAL COMMERCIAL	3307	3280	2952	3175	14,575	15,690	16,130,273	20,184,983	253,500	274,880	1055	1125	400	430	1945	1870	16,408,007	20,485,433

volume growth will catch up with the decline in unit price and therefore diodes, in dollars, will show a 17% increase. Ion lasers, which
are strong across several markets, should post a 14% increase.
Excimers, we expect to see pick up to a 13% increase. Dye lasers,
which have been strong in both R&D and in some of the therapeutic
medical areas, up 12%. Solid state should stay rather lackluster at 5%
growth. He-Ne, because of its conflict with diode in some markets,
could end up flat. If our projections in some of these principal
markets are right, the implication would be a decline of 11% in the CO_2
marketplace.

      This, then, is the broad economic landscape as best we can measure
it. In the talks that follow, my colleagues at Laser Focus and from
the industry will give you a much more detailed discussion of the
technologies that are both driving as well as reacting to these markets, and the more detailed dynamics within each of these various
important market segments.

AUDIENCE QUESTIONS
Q: WHY ARE DIODE LASERS IN THE BARCODE SCANNING SECTOR INDICATING A
      HIGHER GROWTH RATE IN DOLLARS THAN IN UNITS, I.E., A GROWTH IN
      PRICE?

A: Consumers may be willing to pay more for diode technology. Our
speaker on barcode scanning can address that question later.

EMERGING LASER TECHNOLOGIES
IMPACT ON THE MARKET IN 1988

Lewis Holmes
Executive Editor, Laser Focus Magazine

The outline of my talk involves three primary areas. First, an overview of current trends in the commercial laser scene. I want, in connection with that, to point to the fastest developing technologies areas. And finally to mention some of the technology as I see it that are on the horizon. I'm going to try to provide some sort of trend line of what's going on now, but I would be remiss if I didn't attempt to give you a broad view of what is going on across this whole spectrum of laser types.

I would like to start with the viewgraph (Fig. 1) which summarizes areas that, in my judgment, are perhaps the fastest developing at present. First of all, there is the whole area of diode lasers. Diode lasers are becoming a kind of basis technology, enabling technology for other lasers within the industry. They are terribly important, they are adopting a role that is somewhat similar to the role, I suppose it is not quite as extreme in the laser industry by any means, but they are obtaining a somewhat similar role to say, integrated circuitry within the semi-conductor industry, which when it burst on the scene in the 1970's impacted everything in electrical engineering. Diode lasers have not progressed that far and undoubtedly never will in our industry, but they are impacting many other types of lasers, and that impact is becoming very, very clear at present. Not only that, the lasers in themselves have applications which are terribly important. We have a particular interest in the diode laser area itself in the development of higher powers and shorter wavelengths. This is particularly important in devices that are used in the large-scale commercial applications and in devices that are used for pumping solid state lasers which is the second area picked out on this viewgraph. Here we are seeing the development of more power, of new wavelength capability, new source materials.

The third area on this overview graph is excimer lasers. Excimer lasers are about a decade old now, and they are moving away from the applied research stage and into industrial applications. There is tremendous ferment of activity in the excimer laser area, and although there are only a few major players in that area, new companies are jumping in and new applications are opening up. And new technology is being developed.

Finally, I have listed the color shifting techniques. Color shifters are not lasers, but they are used with lasers in order to access different wavelength regions. This is something which is broadening the technological capabilities of lasers and is having a very strong impact on technologies and product offerings.

Most of the material that I am going to be presenting is based on a survey that we did at the end of 1987, and a number of you in the audience I recognize as having to contributed to that survey. I hope that the things that I say relating to your own particular areas don't

Figure 1

Areas of Most Intense Development Activity

DIODE LASERS
    HIGH POWER
    SHORT WAVELENGTH

DIODE-PUMPED SOLID-STATE LASERS
    MORE POWER
    NEW COLORS (BLUE; OTHER VISIBLE AND IR WAVELENGTHS)

EXCIMER LASERS
    INDUSTRIAL ADAPTATION
    NEW MODELS (HIGH PRF AND ENERGY; OSC/AMPL; F-O CONTROL)

COLOR-SHIFTING TECHNIQUES
    NEW NONLINEAR OPTICAL CRYSTALS
    NEW LASER-RELATED OPTIONS

hold any surprises for you, because then I will have made a mistake.

I am grateful for your cooperation, and as Gary Forrest said, one of the services that we can provide at Laser Focus is to act as kind of a clearing house and synthesizer of information that we gather from you people. I would like to say that there are other areas that are developing very rapidly, and certain people would put other technologies down on that list. There has been a lot of work in ion lasers which is surprising for a mature technology. There has also been work in lamp-pumped solid state lasers, which is really quite impressive.

Another point that I would like to make right now is that in spite of any economic uncertainties that we may be facing, technology watchers like myself look forward to another 1/2 year or so with dead certainty that we will see a lot of new technical capabilities being introduced in lasers. The reason for that is that the R&D that goes into the products that are going to be introduced in April of this year is essentially all done now. So, for a certain amount of time, whatever 1988 brings, we can be certain that it will bring new technological developments. In connection with that I would like to read from a report that I wrote in December 1987, which did not appear in print in quite this form, but I think illustrates the dynamism of this industry at the present time. "For the near term, laser industry sources indicate continued rapid product development. During a recent series of interviews, these sources pointed to planned 1988 introductions of new solid state lasers, diode lasers, short pulse dye lasers, low and medium power CO_2 lasers, excimer lasers, nitrogen pumped dye lasers, He-Ne lasers, ion lasers, He-Cd lasers and tunable solid state lasers." In other words you can expect to see continued developments in just the whole broad spectrum of laser types.

DIODE LASERS

What I am going to do is walk through the different graphs that you have in your summary. We are starting with high powered diode lasers here, which I described previously as a basis technology for certain other laser types. What we are seeing is the availability of these devices from more suppliers, and the movement to higher power ratings. This technology was pioneered commercially in the U.S. by Spectra Diode Labs which offers a phased array device. 1987 saw Sony jump into the picture with a device that has comparable apertures on the beam coming out but it is actually a broad stripe device. The power ratings have reached 1 watt CW now, which makes them quite substantial sources of energy. There is intense work on the development of high powered diode lasers for use in applications such as pumping of solid state lasers at many different companies. You may see new supplies coming from unexpected places in 1988.

With diode pumped solid state lasers, this is an amazing story of commercialization of technology. Three years ago or so these were not a product, they were kind of a R&D dream. In that period of time we have seen emergence of a half dozen or so suppliers; there are 30 models of diode pumped solid state lasers that you can buy. You can get infrared wavelengths, frequency doubles green wavelength. You can buy gain switched and Q-switched versions. You can buy narrow band CW versions. Just a tremendous variety in this technology. The point is frequently made also that because these devices are small, you can use small crystals in them, as the laser medium, and thereby provide access to new and attractive types of laser crystals that haven't been usable in the past.

We can expect to see power levels increase. The current IR maximum is 3 watts now. A company called Laser Diode Products in Earth City, Missouri, has just announced a 3 watt device and they have announced that they will introduce a 5 watt device in the first quarter of 1988. When you frequency double using KTP, you can get green output and that typically is down below 100 milliwatts. There is quite a bit of variety in the technology as to whether you feed the optical pumping power in through an optical fiber as Spectra Physics does or if you have electrical connections as the other companies have.

By first industrial applications, I am referring to the fact that ESI has cooperated with one of the suppliers and is putting a laser of this type into one of its semi-conductor processing systems.

Moving on, we get to other short wavelength diode lasers. Moe Levitt has already mentioned a good deal about this. The wavelength is being driven down on these devices, and that is an important factor from several points of view. But in terms of the application to optical memory systems, the shorter the wavelength the higher the packing density on your storage medium. So there is a real driving force there to get the wavelength down. There is a recent Electronics Letter talking about CW operation of the diode laser at 640 nanometers. He-Ne lasers are 633 nanometers, so you are getting very, very close there. And we are predicting that visible red diode lasers will be launched on the market in this coming year. There was information on this presented by Sony and by NEC in yesterday's conference meeting here, which our chairman today also chaired.

High frequency capabilities refers to the driving of modulation capabilities higher in devices of a different kind that are used in data communications, where there has been a trend towards getting into the microwave regime in order to transmit information faster. Long wavelength sources are used in fiberoptic communications in the tele-communications industry. And here we have a movement toward the extreme narrow line width that distributed feedback designs provide. There is a movement in that industry toward developing coherent trans-mission systems for which narrow linewidth sources are very favorable.

Just to mention a couple of things under tunable diodes and solid state lasers. They don't really fit too closely with the rest of this chart but, they are an important area. I would like to mention that in the color center laser area, this is kind of a niche product area, but there is development here as well. There is a new color center laser on the market which is actually made out of table salt, sodium chlo-ride. When this was announce by Burleigh last fall, I was tempted to put in the headline "The First Edible Commercial Laser", which Dave Farrell is happy I didn't do. That provides up to something like a third of a watt in the 1.7 - 1.9 micrometer wavelength region.

SOLID STATE LASERS

We are moving on to other types of solid state lasers. Some are bigger ones. For science applications we see a kind of ripple effect of the diode laser pumping technology. The narrow linewidth diode-pumped solid state lasers can be used to seed larger pulsed Q-switched neodymium YAG lasers. The result is a tremendous improvement in tem-poral smoothness and in reproducibility of the pulses. This has a great positive effect on the kinds of experiments, the kinds of science experiments, that you can do. I'd love to mention, incidentally, that much of the material on the diode lasers and the diode-pumped solid state lasers you can find in our publication, (1988 Annual Review of Commercial Laser Technology), which reprints a number of key articles from Laser Focus magazine last year. Gary Forrest did a number of very good articles talking about those particular technologies.

In the CW pumped devices, we are seeing an increase in the Gaussian-mode output that is available. The state-of-the-art is now above 20 watt CW and about 2 watts in the green from these devices. Those devices are often mode locked and used to sync-pump dye laser systems. Also, in this last year, Quantronix introduced an alternative mode-locked source for that purpose which gives shorter pulses out and intrinsically higher thermal stability than you could get from the neodymium YAG. The standard pulse length on that is 35 picoseconds in the green.

In the industrial area, the high power neodymium YAG lasers, we have seen the introduction of a number of kilowatt class devices from Japan. That is very striking. These are multi-mode outputs, one kilowatt is a heck of a lot of laser energy. Of course, when you start talking about that kind of power level, you talk about competing with CO_2 lasers, and the question is what you can do with this 1 kilowatt output. I know Dave Belforte has told me that there is work needed to document the effects of the beam at this power level. There is no sign yet of the industrial slab laser. Journalists have been predicting industrial slab lasers for 15 years now. The initial patent runs out

in 1989 I think. I'm not going to predict that a commercial slab laser
is going to arrive this year. I have already been caught at least once
on that. This is something that had been developed at General Electric
in Schenectady, New York, which holds the basic patent on it. It is
called by various names: zigzag slab laser, face-pumped total internal
reflection laser. The advantage is that it produces very good beam
quality at potentially high power levels. But it hasn't been industrialized except, as I understand it, in a material processing system
that Moshe Lubin developed at Hampshire Instruments where it is used to
generate laser power that produces x-rays for lithography. There is
continued research on this, and there is a big project in Lawrence
Livermore in high-average-power studies.

DYES AND EXCIMERS

Ultrashort pulse dye lasers have benefited from the improved CW
pumped sources that I mentioned just above. There are new pulse ampli-
fication concepts being developed. A recent example is the use of
copper vapor lasers to pump dyes in order to amplify pulses, that is
one thing that is around now. Another thing that I would mention is
that a company in the Rochester, New York, area has come up with a kind
of novel scheme for marketing these things. The company, Clark Instrumentation, is an offshoot of Newport Corporation. It is going to sell
kits of mounts and mirrors and dye jets, etc., so that you can build
your own femtosecond dye laser.

In the flashpumped dye laser area, applications have been the big
news. They have FDA approval for a couple of treatments, including
lithotripsy which is shattering of stones, something that you might
normally use ultrasound for, but there are certain positions in the
body where you can't get the sound waves to, but you can feed a laser
pulse in through a fiber, and shatter the stones that way.

On the final thing here, the Littman short-cavity design of
single-mode and longitudinal-mode laser pumped dye laser is being
commercialized by Lumonics, that is, I think, an important advance. It
makes a long single mode tuning range available.

In the excimer laser area, we have seen a growth of industrial
applications. Excimer lasers are used for marking and lithography.
There has been an expansion in medical research where they are used in
sculpting the cornea in order to correct for vision defects. They are
available in a number of new technologies including the use of x-ray
preionization which provides uniform preionization over a large volume,
allowing you to get bigger pulses out. Long pulses have been introduced in the last year, and oscillator/amplifier sets have been introduced. Special industrialized lasers have been introduced. The
largest supplier, Lambda Physik, introduced a multiplexed local area
network to control the devices with distributed intelligence on the
different parts of the lasers. This is in their standard line of
lasers. There have been a number of new suppliers coming onto the
market. In 1988 you can look for a new waveguide sealed excimer laser
which is microwave excited, coming from Paul Christiansen of the
University of Maryland who has formed a company called Potomac
Photonics to market it.

GAS LASERS

Ion lasers have seen a 25% increase in the power -- UV power increasing to 5 watts. This is tremendously important in terms of various applications. I was just speaking to somebody in England who used the 5 watts of UV in order to get subpicosecond pulses in the blue from a dye laser, something he said he could never do before this last year with the availability of higher power. A number of different tube design improvements, and a number of different concepts are being introduced in order to access this continually important market.

He-Ne lasers have been somewhat steady in the last year with evolutionary improvements in the technology. The big news is that Spectra Physics sold its high volume line to Uniphase and has changed the structuring of that industry.

CO_2 lasers - we are seeing more power out of sealed off devices; there are going to be a couple of 100+ watt sealed off devices coming out in the next month or two. In the higher powered CO_2's for industrial applications in the 500- to 1000-watt range, there is also a good deal of ferment here which is kind of interesting in view of the flatness of that market. We've seen important new concepts in that area coming from abroad, coming from the Federal Republic of Germany, and there may be additional surprises in the offing from this country as well.

Copper vapor lasers have higher power, up to 60 watts; actually there was a talk yesterday about a 100 watt device that is now commercially available. It is a beautiful marketing ploy, you actually buy three of the company's lasers and stack them together to get the 100 watts. And there are air-cooled options available.

In Japan a white-light laser has been introduced, which is of interest for film writing. This has four lines in the visible.

Chemical lasers. There are low powered designs and there are high powered up to 500 watts that are now available from the only supplier that I know of, Helios. I believe TRW just tested a 2 megawatt chemical laser for SDI out here in California a week or so ago. There is quite a bit of potential in that area.

FUTURE TECHNOLOGY

Finally, I want to go to technologies on the horizon. One of them is surface emitting diode lasers. If you think about it, it makes a lot more sense. If you have a flat chip and have the light come out normal to the surface, it is much more useful if you want to get a lot of light out. That is something that is being worked on at Lincoln Laboratories, at various university laboratories in this country and Japan. It is something that I am sure will come to pass, and you can do it either by etching mirrors into the surface that reflect the light out or by re-designing the laser cavity within the semiconductor diode so that the light comes perpendicularly out.

Color selectable all solid state lasers. Here what I am referring to is a situation in which you or say I as a customer decide for some reason I want a flashlight-size 359 nanometer laser. I call up a company and say, please supply me one of those. The person who is taking orders says, okay - give me your credit card number, and I'll send it out in 8 weeks. I think that kind of approach is on the horizon, that we are going to have the option of picking a color that we

want and being able to order it. You can do this with various
frequency shifting and combining techniques.

    <u>Free electron lasers</u> are on the list. I have to admit, primarily,
because everybody else always puts it on the list. I don't feel very
confident about that area. There are a couple of companies that have
been formed in order to market free electron lasers. They are both
aiming for the medical and the R&D markets.

    I was speaking to Charlie Brau at Los Alamos about this the other
day and asking him about the applications. He said that there are
certain high volume semi-conductor processing applications that are
potentially possible and some of the medical applications look
interesting. I said, "what about the size of these things, can you put
them in a room?" He answered "Not in a usual room." You have to build
big concrete walls to shield from RF, or he recommended digging a big
hole and dropping it in. Also, you have to have a separate control
room and RF supply. There are possibilities of miniaturizing, you can
go to Van de Graaf generators to accelerate the electrons, you can use
something called a microtron. There have been three experiments using
a microtron, one in the U.S., one in Europe, one in the Soviet Union,
and none of them have succeeded. But a lot of people tried to fly
before the Wright Brothers succeeded, so it may come.

APPENDIX:
NEW COMMERCIAL LASERS

Lewis Holmes
Executive Editor, Laser Focus Magazine

During the past year, technical progress led to significant growth in the performance levels of several types of commercial lasers. In addition, virtually all segments of the laser industry recorded at least some improvements in the technology and in its applications.

For the near term, laser-industry sources indicate continued rapid product development. Highlights of 1987 activities and examples of planned 1988 product launches make up the discussion that follows.

HIGH-POWER DIODES AND DIODE-PUMPED ND:YAGS

Since 1985, a broad development current has carried diode-pumped solid-state lasers into the mainstream of commercial laser technology (see the November 1987 LF/E-O, p. 62). Involving ongoing work on diodes, solid-state lasers, and nonlinear optics, this current continues to bubble with activity.

The power output from high-power diode lasers has increased steadily since Spectra Diode Labs, San Jose, Calif., introduced the first high-power phased-array diode lasers in 1984. Based on GaAlAs semiconductors, such lasers emit in the infrared (IR) at a wavelength near 800 nm.

Last year, both Spectra Diode Labs and Sony Corp. of America, Cypress, Calif., released GaAlAs diode lasers with unprecedented continuous-wave (CW) power ratings of 1 W (see the September 1987 LF/E-O, p. 76). These are single-chip devices. Laser Diode Products, Earth City, Mo., announced a 1-W CW diode-laser package comprising two bars of emitter elements.

The Spectra Diode Labs and Laser Diode Products packages contain thermoelectric coolers. Regulation of the temperature keeps the diode wavelength constant in applications such as neodymium-laser pumping.

At present, a half-dozen manufacturers have developed and sell solid-state lasers pumped by these high-power diodes. Manufacturers of diode-pumped lasers include Adlas, Luebeck, F.R.G. (represented in the U.S. by AB Lasers, Concord, Mass.); Amoco Laser Co., Naperville, Ill.; Laser Diode Products, Earth City, Mo.; Lightwave Electronics, Mountain View, Calif.; and Spectra Physics, Mountain View, Calif. Some 30 available models include CW, Q-switched, and frequency-doubled (green) versions. Many are portable flashlight-size units. Spectra-Physics minimizes head size by delivering pump power fiberoptically.

The usual host crystal in these solid-state lasers is Nd:YAG. An alternative crystal, Nd:YLF, offers higher Q-switched energies under certain conditions.

Spectra-Physics has noted that working with small laser crystals is an attraction of diode-pumping technology. Crystals that do not readily grow to large sizes can be utilized. The company, which already makes a wide range of diode-pumped Nd:YAG and Nd:YLF lasers, is investigating novel systems to produce different wavelengths or achieve other favorable characteristics.

Lightwave Electronics has exploited the versatility of this young technology. Established in 1986, the company already manufactures a diverse range of diode-pumped lasers. They include CW injection seeders, narrow-linewidth single-mode ring lasers, and Q-switched lasers. Wavelengths include three IR lines, 1.06 um and 1.3 um from Nd:YAG and 1.05 um from Nd:YLF, and green.

New products appear almost monthly. Adlas, with IR and green models, recently introduced a 1.3 um product, which has interest for fiberoptic testing and for ranging. In 1988, the company expects to introduce other new Q-switched lasers and frequency-doubled pulsed lasers.

Amoco Laser, which started up in 1987, has shipped lasers at two IR wavelengths and has announced a green product. This company has focused on the compactness, wavelength versatility, and suitability for high-volume manufacture of diode-pumped lasers.

Different wavelengths come from frequency doubling, or by mixing frequencies from the diode and solid state lasers, in nonlinear optical crystals such as KTP. IBM recently described the generation of blue laser light by the nonlinear intracavity mixing of IR diode- and Nd:YAG- laser lines. Shortly thereafter, Amoco Laser demonstrated a similar result, which the company says will appear in a product in 1989.

Amoco uses two diodes--one to pump the Nd:YAG and the second for mixing with the Nd:YAG output in KTP. According to Amoco, sum- and difference-frequency techniques ultimately will access a wide range of wavelengths in the visible and IR. A strong advantage of the mixing approach is that the output intensity varies directly with diode current at very high modulation rates.

VISIBLE DIODE LASERS

Not all parties agree that diode-pumped solid-state lasers will make ideal visible lasers. Other work aims to bring diode-laser output directly into the visible.

In 1986, Matsushita Electric, Osaka, Japan, demonstrated frequency-doubling to the blue IR diode-laser output (see the September 1986 LF/E-O, p. 8). To boost conversion efficiency at the low light levels available from a diode, doubling takes place in a nonlinear optical waveguide (see the August 1987 LF/E-O, p. 30). Although beam quality suffers in the process, waveguide doubling represents a long-term option for diode lasers.

At the opposite end of the visible spectrum, design changes such as increasing the aluminum concentration or forming multiple quantum wells have driven diode-laser wavelengths to well below 700 nm in laboratory devices. Several Japanese companies including Sony Corp. and NEC, both in Tokyo, are said to be working toward the commercial introduction of red diode lasers in 1988.

OTHER FIXED-FREQUENCY DIODE LASERS

Large-volume production of GaAlAs diode lasers concentrates on compact-disk and related applications. Discussion of this important segment of laser technology forms part of a review of optical-memory developments to appear in the February 1988 LF/E-O.

Longer-wavelength InGaAlAs diode lasers serve as fiberoptic transmitters at 1.3- and 1.55- um wavelengths. A half-dozen firms have launched narrow-linewidth InGaAlAs based on distributed-feedback designs, to provide for high-speed communications (see the December 1987 LF/E-O, p. 142).

The April 1988 LF/E-O will overview developments in InGaAsAs diode lasers for fiberoptic applications.

LAMP-PUMPED SOLID-STATE LASERS

Diode-pumped lasers serve in an important capacity as an add-on to a more conventional laser system. The Lightwave Electronics diode-pumped injection seeder provides a narrow Nd:YAG line that smooths and narrows the spectrum of high-energy pulses from flashlamp-pumped and Q-switched Nd:YAG lasers.

Both Spectra-Physics and Quantel International, Santa Clara, Calif., market the Lightwave Electronics seeder with their pulsed Nd:YAG lasers. According to Spectra-Physics, this product line, which was introduced in 1986, "took off" in 1987. Quantel just introduced a new injection-seeded, pulsed-Nd;YAG system that emits in TEM_{00} spatial profile. Frequency-doubled models provide energies of up to 0.5 J in 4- to 5-ns pulses.

In the last quarter of 1987, Lumonics, Kanata, Canada, introduced a new high-energy Q-switched Nd:YAG laser. Configured as an oscillator/amplifier, the Model HY1200 emits 1.2-J pulses in the IR, which frequency-double to 0.5-J pluses in the green. The company has undisclosed plans for additional new scientific Nd:YAG models in 1988.

In 1987, both Lumonics and Quantel introduced new compact pulsed Nd:YAG lasers. The Lumonics HY400 provides 0.4-J pulses at a 10-Hz pulse-repetition frequency (PRF). The Quantel YG660-A was engineered for light weight and low cost. It operates at 10- to 50 Hz PRF, with IR pulse energy of about 0.3 J.

Continuously pumped solid-state lasers reached new capabilities in 1987. Coherent entered the scientific Nd:YAG market with the highest CW power available. The company's new Antares model emits up to 25 W of Gaussian-mode output at 1.06 um. Mode-locked, it delivers 100-ps IR pulses, and the output frequency-doubles to 2 W of 70-ps pulses in the green.

Spectron Laser Systems, Rugby, U.K., made the first deliveries of its 20-W CW Nd:YAG laser in 1987. Pumped by a single arc lamp, in contrast to symmetrical two-lamp pumping in Coherent's Antares, the Spectron laser nevertheless also produces up to 2 W of green.

Quantronix Corp., Smithtown, N.Y., introduced an Nd:YLF laser in 1987 and has already announced a 60% increase in its IR power specification to 8 W at 1.05 um. Advantages of this laser, in comparison to Nd:YAG, include higher intrinsic stability and shorter mode-locked pulses. In frequency-doubled mode-locked operation, it provides 35-ps pulses and produces an average power of 1 W.

In the industrial arena, Lumonics introduced a new Nd:YAG product line in 1987. The new 700 Series of lasers, with output power to 400 W, has an "ageless" ceramic cavity with no metallic coating. The company manufactures these lasers in Rugby, U.K. Lumonics Material

Processing Corp., Livonia, Mich., markets them in No. America.

Two Japanese companies have recently pushed power specifications on multimode industrial Nd:YAG lasers to above 1 kW. Toshiba Corp., Kawasaki, specifies 1 kW CW and 1.4 kW in a modulated output mode. NEC, Tokyo, specifies 1.2 kW CW. Fiberoptic power delivery reportedly is possible at these power levels.

The NEC and Toshiba systems operate on Nd:YAG rods. Little, if any, movement appears to have occurred in the long-awaited development of an industrial slab laser, which would utilize designs developed and patented by General Electric, Schenectady, NY.

Quantitative testing must determine the industrial utility of kilowatt Nd:YAG lasers. They could, in principle, displace CO_2 lasers in certain applications. For more on new industrial laser technology, see a forthcoming review in the March 1988 issue of LF/E-O.

TUNABLE SOLID-STATE AND DIODE LASERS

In November 1987, Burleigh Instruments, Fishers, N.Y., announced a tunable NaCl F-center laser. Based on developments at Cornell University, the device tunes in the near IR between 1.7 and 1.9 um. Burleigh pumps the laser with Nd:YAG radiation. Tunable output power is as high as 350 W.

Laser Analytics Division of Spectra-Physics, Bedford, Mass., has installed molecular-beam epitaxy (MBE) equipment to improve its tunable lead-salt diode lasers. Product introductions are planned in 1988. Benefits of MBE include higher operating temperature and wider tunability, together with increased frequency stability and improved control of the manufacturing process.

Schwartz Electro-Optics, Concord, Mass., intends to introduce tunable $Co:MgF_2$ lasers before the end of 1988. Pumped by the 1.33 um line of Nd:YAG, these pulsed lasers will tune over the range from 1.8 to 2.5 um.

EXCIMER LASERS

Industrial applications have figured prominently in excimer-laser developments during 1987. In addition, new scientific excimer lasers have appeared, and medical studies based on excimer lasers have intensified. This segment of the laser industry continues to be very active.

During 1987, Lambda Physik, Goettingen, F.R.G., released a series of new excimer-laser products. They include a 2-J device, the highest-energy excimer laser commercially available; a long-pulse device; and a fluorine-optimized, high PRF laser. The company's U.S. headquarters are in Acton, Mass.

The Lambda Physik high-energy Model 401 incorporates x-ray preionization. That allows a large uniform spatial profile to be achieved across a beam section measuring approximately 4 cm x 5 cm. Pulse energies are up to 2 J with XeCl operation at 308 nm. PRFs extend to 100 Hz.

In the Model EMG 602, Lambda Physik has stretched the pulse length at 308 nm by one order of magnitude to 250 ns. Lengthened pulses have lower peak powers, simplifying fiberoptic transmission.

In December, the fluorine-optimized Model EMG201F appeared. It increases KrF pulse energy to 0.6 J at 150-Hz PRF, providing 90 W of output power at 248 nm.

At the 1987 Conference on Lasers and Electro-Optics (CLEO), Questek demonstrated a marketable version of its "power oscillator/-power amplifier (POPA)" concept. A new synchronizer and delay generator, Model 9200, controls the timing on a pair of Series 2000 lasers in tandem. Specified POPA outputs range up to 1 J at 100-Hz PRF when the units are operating on KrF at 248-nm wavelength.

The Model 9200 also can control a third excimer laser, possibly pumping a dye, to aid in carrying out two-color pump/probe studies. The device synchronizes to _1 ns. Optical fibers carry light for timing purposes from the laser discharges to the controller.

During the past one to two years, Lumonics has devoted most of its research and development effort in Kanata to industrial excimer lasers. Following a big sale to an undisclosed customer, the company demonstrated a serial version of an industrialized laser. This laser, Model Index 200, is rugged and has a conservative power rating of 40 W at 248 nm. Lumonics showed it ablating polymer film at CLEO in an associated workstation Model MPS 100.

XMR, Santa Clara, Calif., delivered 150-W industrial excimer lasers in 1987. The company announced a lower-power version of this device.

In 1987, new companies entered the excimer arena. In Munich, F.R.G., Siemens announced that it would make an industrial excimer laser. The Japanese company NEC, Tokyo, released its first excimer laser, a 10-W XeCl device.

Lambda Physik is starting off the new year with a major technology introduction. In January 1988, the company introduces its "5th generation" excimer laser, the LPX, which will replace the standard EMG100 series.

Most striking about the LPX is the means of control. The control circuits are based on a fiberoptic local area network (LAN). Key modules such as the energy monitor and high-voltage power supply connect to the LAN. They are equipped with microprocessors that receive and interpret control signals transmitted at a 400-Mbit/s rate.

The LPX laser comes with a portable minicontroller keypad. The control circuits interface readily with an IBM PC.

Other features include a modular power supply and modular head construction.

ION LASERS

For a mature technology, ion lasers showed considerable change in 1987. Maximum power levels increased 25% to 25 W in the visible and more than 25% to 5 W in the UV. Low-price argon and krypton designs hit the marketplace, while modular concepts surfaced in several companies' product lines.

Coherent and Spectra-Physics vied strongly with one another at the high-power end of the technology. Spectra-Physics introduced its first large-frame metal-ceramic argon-ion lasers. Model 2030 supplies 20 W of visible and 1.5 W of output in the usual multiline UV region (351-364 nm). Corresponding specifications on UV-optimized Model 2035 are 15 W and 3.5 W. This laser also provides 400 mW in the deep UV (275-305 nm).

Coherent's initial 1987 contender, the Innova 100-UVE, appeared in April with 20-W visible and 4.0-W UV specifications. The company also specifies 9.0 W in the single 514.5 nm visible line.

In December 1987, Coherent raised the visible multiline specification on the Innova 100 to 25 W. Model Innova 100-UVE25/5 now offers 25 W in the visible and 5.0 W in the UV.

Each company claims improvements in tube lifetime due to control of manufacturing environments and progress with components such as lower-absorption quartz Brewster windows. Such changes also have helped to make the higher power ratings possible.

Spectra-Physics brought out a compact argon laser, Model 2016, which is smaller and less expensive than the comparable Model 2020. It provides multiline or single-line visible outputs.

NEC announced an argon-ion laser based on aluminum-nitrite tube technology. The first of a planned full line of such devices, the Model GLG 3400 emits 4 W in the visible.

Several companies including Continental Laser Corp., Mountain View, Calif.; Cooper LaserSonics, Fremont, Calif.; and Laser Ionics, Orlando, Fla.; introduced new modular concepts in ion lasers. Continental stacks tubes, with individual single-phase power supplies, in series. The company will extend the capability of such tandem arrangements to 16 W in 1988. Cooper offers an upgradeable 3-W laser, the new Excel 3000, with modular power supply. Laser Ionics has designed its new Series 1500 around a quick-change replaceable tube cartridge.

In contrast to the water-cooled lasers described above, milliwatt argon-ion lasers often appear in internal-mirror designs with air cooling. According to Cyonics, Sunnyvale, Calif., this technology is moving toward higher power for improved signal/noise ratios and higher throughput in commercial applications.

ULTRASHORT-PULSE DYE-LASER SYSTEMS

Coherent's new Nd:YAG laser Antares serves as a pump laser in the company's subpicosecond dye-laser (see the May 1987 LF/E-O, p. 40). In January 1988, the company will reportedly announce a new collaborative venture with Quantel International in ultrashort pulses. The venture will join Coherent's mode-locked CW Nd:YAG and synchronously pumped dye-laser technologies with Quantel's pulsed Nd:YAG and dye amplifiers. The combined system will yield tunable subpicosecond output pulses at millijoule energy levels and 50-Hz PRF.

Also in January 1988, Clark Instrumentation, a new company in Pittsford, N.Y., will start to offer femtosecond-dye-laser kits.

NITROGEN LASERS AND NITROGEN-PUMPED DYE LASERS

Laser Science Inc., Cambridge, Mass., introduced a nearly diffraction-limited nitrogen laser, Model 337ND. It emits 200 uJ, 3-ns pulses at 20-Hz PRF for diagnostic applications and dye-laser pumping.

Pumped by the 337ND, the Laser Science DLMS dye laser covers wavelengths between 360 and 900 nm. Output energy from the grating-tuned device is up to 50 uJ. A grazing-incidence version of this laser has 0.01-nm linewidth.

Photon Technology International, Princeton, N.J., introduced its first pulsed nitrogen laser, Model 2300. This device reportedly is made by former employees of PRA International in London, Ontario, Canada. According to Photon Technology, the Model 2300 resembles a big nitrogen laser made by PRA, but has an increased energy per pulse of 1.4 mJ, as a result of a head redesign. In 1988, Photon Technology will release two mating dye modules, the PL201 and the high-resolution PL202.

The former PRA International laser operation has become PRA Lasers Inc., London, Ontario, Canada, part of Laser Photonics, Orlando, Fla. PRA Lasers still manufactures the previous range of high-peak-power nitrogen lasers and nitrogen-pumped dye lasers. Laser Photonics reportedly also has brought high-energy nitrogen-pumped dye-laser technology from Cooper LaserSonics. That move gives Laser Photonics access to a broad range of pulsed dye-laser capability.

OTHER PULSED DYE LASERS

The energy output from flashlamp-pumped dye lasers increased in 1987. Candela Laser Corp., Wayland, Mass., introduced the SLL-8000 that emits up to 40 J in 2.5 us pulses. The company has delivered a special 100-J device to Sandia Laboratories for use in lithium ioniza-tion for particle-beam fusion experiments.

Nd:YAG- and excimer-pumped dye lasers also stepped forward in 1987. Lumonics brought out a short-cavity pulsed-dye-laser design. The Hyper-dye-SLM provides smooth temporal performance from single-longitudinal-mode operation. The time-averaged linewidth of output 2-ns pulses is very narrow at 500 MHz. In addition, a carefully con-figured synchronous-scanning mechanism tunes the wavelength over a range of several wave numbers without mode-hopping.

Recently, Lambda Physik achieved single-frequency operation in its FL3002 dye laser by equipping it with a high-finesse etalon and pumping it with 50-ns excimer pulses. Details will be presented at the January 1988 O-E/LASE.

HE-NE LASERS

A review in the December 1987 LF/E-O, p. 60, showed suppliers of 633-nm He-Ne lasers moving toward industrial restructuring. Uniphase, Sunnyvale, Calif., is taking over Spectra-Physics low-power He-Ne manufacturing line.

At least four companies now make green He-Ne lasers with 543-nm wavelength: Melles Griot, San Marcos, Calif.; PMS Electro-Optics, Boulder, Colo.; Spindler & Hoyer, Goettingen, F.R.G. and Milford, Mass.; and Siemens, Munich, F.R.G. and Iselin, N.J. At present, the maximum specified green He-Ne power is 0.75 W. Improvements in the optics should make higher powers available in 1988.

At the 1987 CLEO, Spindler & Hoyer showed a frequency-stabilized He-Ne laser with specified frequency instability of 1×10^{-8}. At that time, Uniphase introduced a low-priced stabilized He-Ne laser. It stabilizes electronically on an isotopic neon line.

CO_2 LASERS

The field of low-power CW CO_2 lasers stayed relatively quiet in 1987. Product innovations included a compact 2-W device from Hughes Aircraft, Carlsbad, Calif.; a Stark-cell-stabilized laser from MPB Technologies, Dorval, Quebec, Canada; and a rack-mountable power supply from Laser Photonics.

Edinburgh Instruments, Edinburgh, U.K., which makes RF-excited CO_2-waveguide lasers, reportedly plans to show a new 50-W device at the 1988 CLEO. Quantum Electronics International, Concord, Mass., represents Edinburgh Instruments in the U.S.

At industrial kilowatt power levels, MBB, Munich, F.R.G., introduced an RF-excited fast-axial-flow laser with 1-kW rating. Trumpf, Ditzingen, F.R.G., brought out a 5-kW TLF laser, which has electrodes outside the reaction zone to protect them from erosion. Rofin-Sinar, Hamburg, F.R.G., now part of the Siemens group, scaled up a previous CO_2-laser design to 6 kW.

Industrial laser innovations will be presented in a March 1988 LF/E-O article.

Limited changes have occurred in the technology of pulsed transverse-electric atmospheric-pressure (TEA) CO_2 lasers for marking applications. Lumonics replaced spark gaps by thyratrons, to reduce costs of consumables (plumbed-in air) to customers. A new galvanometer-based scanner speeded character-marking rates to 100 per second. This new high-speed capability has opened up new applications in electronics.

For lidar and Doppler-shift applications, Laser Science International introduced a pulsed TEA CO_2 laser with automatic frequency control. A CW local oscillator actively stabilizes the Model PRF150AFC. Line-tunable output pulses from this laser deliver energies of up to 200 mJ at PRFs to 150 Hz.

COPPER-VAPOR LASERS

In 1987, Cooper LaserSonics, Pleasanton, Calif., launched the 2000 Series of copper-vapor lasers (CVLs). These CVLs replace earlier models, except for the highest-power, 40-W device. In comparison with earlier Cooper products, they provide lower RF noise and higher short- and long-term stability. The new lasers utilize compact switching power supplies.

Oxford Lasers, Oxford, U.K., introduced a 60-W CVL, Model CU60, in 1987. This laser has the highest commercial power specification.

The company additionally increased the power rating on its range of air-cooled CVLs to a new high of 15 W. These transportable devices operate from single-phase power supplies.

HE-CD LASERS

At the 1987 CLEO, Omnichrome, Chino, Calif., demonstrated a life-prolonging modification on its He-Cd tubes. A heatable condensor section near the end mirror allows for boiling away of cadmium deposits. Lifetime improvements of up to 33% are claimed for the innovation at no extra charge to purchasers.

Also at that CLEO, Liconix, Sunnyvale, Calif., demonstrated operation simultaneously at blue (442-nm) and UV (325-nm) wavelengths. A prism separates the two outputs if needed.

In Japan, NDK Laser, Tokyo, showed a 15-mW white-light He-Cd laser at InterOpto 87. With five visible wavelengths, this laser could find use in direct-film-writing applications.

Among product innovations projected by Liconix for 1988 is higher UV power. Taking advantage of improvements in UV ion-laser optics, the company intends to launch a 30- to 40-mW UV He-Cd laser before the end of the year.

CHEMICAL LASERS

Helios, Longmont, Colo., the U.S. manufacturer of HF/DF chemical lasers, designed its new Minilaser for compactness and ease of use. With single-button turn-on and fault-finding diagnostics, this laser suits requirements of medical or fiberoptics researchers. Single-line output powers at wavelengths near 2.5 um, where fluoride fibers transmit effectively, reach hundreds of milliwatts.

In 1987, Helios also extended its laboratory lasers to high-power operation. Multiline IR outputs to 500 W serve in scientific research and material-processing studies.

ADDITIONAL INFORMATION

LF/E-O Laser of the Month articles in 1987 have discussed many of the developments noted above. Detailed listings of laser companies and their products appear in the Laser Focus 1988 Buyers' Guide.

LASER OPTICS MARKETPLACE
WHO'S MAKING THE PROFITS

David Kales
Business Editor, Laser Focus Magazine

The intracavity optics market continues to keep pace with the laser industry. And like the laser industry, as it matures, its growth is slowing in all segments. Intracavity, or internal optics, include mirrors, prisms, beam splitters and Brewster windows. The optics external to the cavity are called beam handling optics, and include laser system optics for medical, industrial, entertainment and scientific applications. They make up 125 million dollars of the total laser optics market. The internal optics are about a 37 million dollar market at this time.

In 1988, the intracavity laser optics market will increase 11% in units and 8% in dollars. This smaller dollar increase is due to the slight decrease in prices in the He-Ne optics area, which accounts for the majority of the unit volume. The He-Ne optics segment continues to grow in both total dollars and units. The predicted downturn has not yet started to show, and with new product introductions, might even take longer than projected. He-Ne manufacturers having rather large volume requirements, and a willingness to change from standard, expensive optics design, have had the best opportunity to demand a price reduction in optical laser components. Ion laser optics continue to hold their price for large frame lasers, but are harder to manufacture than they were three to four years ago. Absorption requirements for optical components have become more stringent. Small ion lasers are starting to be manufactured in reasonable volume and prices are dropping. With new applications being found in the laser disc/mastering market, one would expect the volume to increase. What is happening, however, is that that the new tube designs are lasting much longer and reducing the tube replacement requirements. There will be a small increase in both volume and dollars during 1988.

Dye laser optics will experience the smallest growth, mainly due to the large inventories and the continued use of dye lasers in the medical field. Almost no increase in optics volume is forecast, while just a slight increase in dollars due to price increases is expected. But demand for CO_2 laser optics took a large drop early in 1987, owing to the downturn in the industrial market both here and in Japan. However, as larger and more powerful industrial lasers are developed, more expensive optics will be required boosting the doll growth in this segment. The small CO_2 laser continues to supply demand for optics as its use in medical and military markets increases.

The largest dollar segment of the market continues to be solid state laser optics, due to the high demand for the expensive laser rod in the medical and military markets. With the arrival of higher powered lasers to the market, continued growth is forecast for the materials processing market segment. The smallest segment of the market remains He-Cd laser optics. Because of the small numbers, however, there is considerably greater growth here than in other segments. Optic size and technical requirements are similar to those for He-Ne lasers. With new processing designs, longer tube life is expected. Longer life should help this laser compete with ion lasers in the laser disc mastering market. The demand most likely will remain small in comparison to other lasers.

Excimer lasers are finding new applications in the medical and industrial markets. Demand for optics is coming from the manufacturers that lead the industry in excimer laser production. Many of the optical components are still used in research and development. As the volume increases, the cost will drop quickly, keeping the overall revenue for suppliers of excimer optics low.

Research laser optics to support the Strategic Defense Initiative, and laser fusion programs, is the greatest opportunity in 1988 for those companies with good test capability. Only a select few will qualify, and not all will make money, as has been the case many times in optics when dealing with government.

Other lasers of interest are metal vapor, carbon monoxide, nitrogen. Demand for optical components for these lasers is still small, but represents 400 thousand to 500 thousand dollars.

THE GLOBAL LASER OPTICS MARKET

Worldwide the laser optics market continues to grow. Currently the overseas market represents about one-third of the total intracavity market. He-Ne, excimer and CO_2 laser optics make up the majority of the products purchased.

This is a breakdown of 1987 and projections for 1988:

Market/Laser Optics	1987	1988
He-Ne	4.6 million	5. million
Ion	7.6 million	8. million
Dye	2.5 million	2.6 million
CO_2	5.2 million	5.5 million
Solid State	9.5 million	10.5 million
Excimer	900. thousand	1.3 million
He-Cd	300. thousand	350. thousand

There is no dramatic change in the proportion of the laser marketplace, solid state still stays in the lead by about the same amount.

Fig. 1 depicts the worldwide market for intracavity, the division between the U.S. and overseas. Again, you don't see too much of a change from '87 to projections for '88.

Fig. 2 shows the beam handling market. This is a much larger market. Companies like Coherent are more in the intracavity, where Melles Griot is a company that specializes more in external cavity optics. One of the things that these figures don't reflect, perhaps, is the tremendous explosion in the semi-conductor lasers, the diode lasers, which do demand tremendous amounts of external optics. There are some big market opportunities in the printers, in optical memories for computers. The Japanese, of course, are very strong in the optics that are going into the CD players, but there is tremendous opportunity for U.S. suppliers in some of these other areas, in external optics.

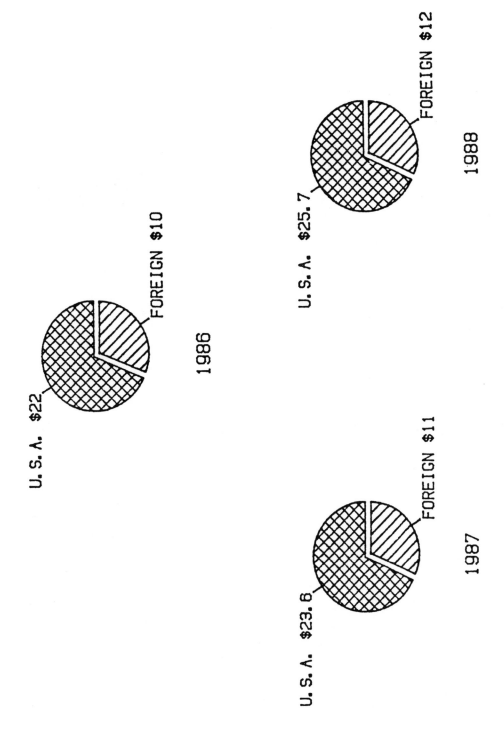

WORLD MARKET FOR INTRA-CAVITY LASER OPTICS
1986 - 1988 OPTICS SALES

FOREIGN $10

U.S.A. $22

1986

FOREIGN $12

U.S.A. $25.7

1988

FOREIGN $11

U.S.A. $23.6

1987

ESTIMATED SALES IN MILLIONS OF $

Figure 1

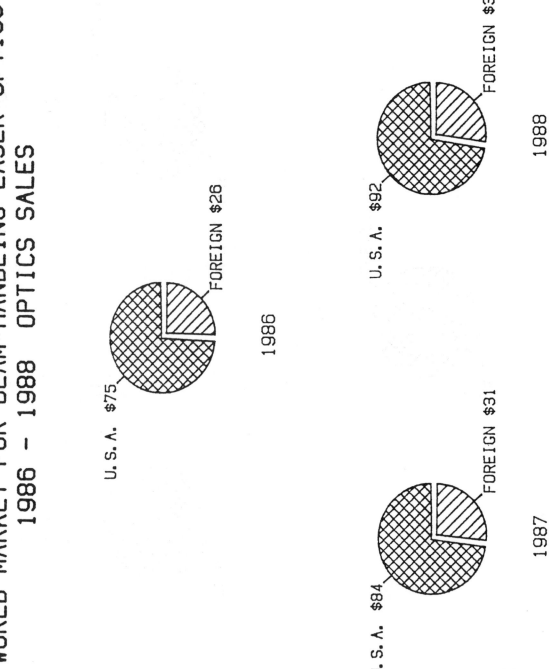

WORLD MARKET FOR BEAM HANDLING LASER OPTICS
1986 - 1988 OPTICS SALES

FOREIGN $26

U.S.A. $75

1986

FOREIGN $35

U.S.A. $92

1988

FOREIGN $31

U.S.A. $84

1987

ESTIMATED SALES IN MILLIONS OF $

Figure 2

Fig. 3 compares the two segments of the laser optics market, beam handling and the cavity optics.

I have some thoughts about 1988. It is a tough year to call. The crystal ball, of course, is clouded with not only the possibility of a recession, but it looks like there are going to be government cutbacks, government funding cutbacks certainly, that are going to impact in the labs and industry. It is going to affect, I think, all aspects of the laser and electro-optics field. But, one thing that I did notice when I did an earlier study on the optics industry -- exports are running ahead for certainly the first part of '87, over '86. And while imports are increasing (I am just talking in terms of optical component products), I think we are going to see tremendous opportunity for the first time in many years for U.S. manufacturers to make headway. And certainly there are tremendous opportunities overseas for suppliers of laser optics. That is the really good news. That is what I would like to leave you with, sort of an upbeat note.

PATENTS AND LICENSES

I am going to switch gears, let me just tell you briefly. I am a reporter. I reported on the cultural revolution in China. I have reported on corporate boardrooms - battles there. U.S. Steel vs the Government back in the early '60's. But I am telling you, after witnessing a Patlex press conference yesterday, the biggest battle I see is going to take place right in the laser and electro-optics area, in your industry. And its impact is going to be certainly very powerful and long reaching. Patlex came in yesterday, (I don't know how many of you people were there), and they said 'The war is over'. And I saw the various people in the industry, CEO's and small companies, patent attornys representing their clients, they were humbled. And they were scared. Patlex threw down the gauntlet ... 'you either get a license for us or we take you to court, and we are going to beat you.' They say it is over, and they haven't tackled the big boys yet, they are going to go after the government because the government is a user and they have a user patent which looks like it is going to be settled in their favor. They are winning all the cases in the courts. Through their four patents, the optically pumped patent, gas discharge, and they have a user patent and they also have a Brewster window patent - they have pretty much covered 85% of all lasers sold. And this user patent is creating all kinds of havoc because they are now starting to go to the users and saying 'hey, I think you ought to pay us a license fee. We have a patent to this.'

So we have a real war starting to emerge. Patlex has, so far, gone after the weaker companies, and they have gained a lot of court precedence this way. But, as a reporter and so called objective observer, it is going to be an exciting time. As far as news goes. As far as the industry goes, I don't know. You can say it is going to have all kinds of ramifications ... survivor of the fittest, obviously. It is going to take a lot of resources. Patlex is a wily company. They have the law on their side right now, but they haven't tackled the big boys yet, and that is going to be fascinating to see. Right now though, it is an incredible situation, and it is going to take the best of management and utilization of management's resources for the individual companies, especially the independent small companies, to remain

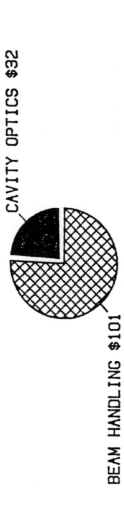

WORLD MARKET FOR LASER OPTICS
1986 - 1988 OPTICS SALES

CAVITY OPTICS $32

BEAM HANDLING $101

1986

CAVITY OPTICS $34.6

BEAM HANDLING $115

1987

CAVITY OPTICS $37.7

BEAM HANDLING $127

1988

ESTIMATED SALES IN MILLIONS OF $

Figure 3

independent, viable and hopefully profitable in the years ahead.

AUDIENCE QUESTIONS

Q: ACCORDING TO THE GRAPHS AND FIGURES, WOULD WE INFER THAT TWO-
      THIRDS OF ALL LASER SALES ARE SOLD IN THE U.S., BECAUSE ONE-THIRD
      IS FOREIGN SALES.

A: No, I would say there is a discrepancy there. I don't think there
is a direct correlation. I am using a little different set of figures
here -- and this is the problem of comparing apples and pears and using
different methodologies, which Larry Giammona has used a little bit
differently. I would not try to make that inference a direct
inference, because laser by laser they vary. That also is just a
general total figure, but by types they vary as far as distribution of
foreign or overseas.

Q: DAVE, YOU MENTIONED THAT ON THE INTRACAVITY LASER OPTICS GROUP,
      CO_2 LASERS SHOWS A GROWTH OF 5.2 TO 5.5 MILLION. YET MOE
      MENTIONED ON THE CO_2 LASERS THEMSELVES A DROP OF 11% FOR 1988.
      COULD YOU TELL ME ABOUT THAT?

Kales: Yes, I am going to comment by being evasive. There are
different people putting together different sets of numbers, and I have
to plead innocent in this case, though maybe I am guilty of using Larry
Giammona's (optics) figures without trying to dovetail them into my
(laser) figures. I do not have a good picture of what is exactly
happening in Japan. And that is the big question mark in my mind, and
in my mind it is making for the discrepancies in the figures. The
figures that I have been using for Japan are really Japan's fiscal year
which ended in March 1987. So a lot has happened since then, and I
just can't give you a direct answer.

Levitt: One factor, and it may not explain this, but one factor that
certainly has to be taken into account in the industrial laser market
at least, is there is a replacement optics market. Therefore, you know
that there is not a direct correlation between unit laser sales and
unit optic sales, there is almost an independent marketing and manufacturing operation for replacements optics which has its own pace and
depends on what is going on in terms of optics lifetimes in job shops
or in industrial plants. I think to begin to get the answer to that
question, we have to unravel how much of this is new optics sales vs
the replacements, among other factors.

SCIENTIFIC LASER MARKET
STILL THE BACKBONE OF THE INDUSTRY

Jon Tomkins
VP and General Manager
Laser Products Division, Spectra-Physics

Good morning. I would like to thank Gary Forrest and Laser Focus for giving me the opportunity to participate in this seminar. After listening to Moe Levitt's overview of the industry outlook for 1988, it is a pleasure for me to represent one of the 'hot' sectors in the market - the old research market. My mission today is to provide a perspective of this market to you as one of the healthier parts of the industry, as one of the backbones of our industry.

In order to accomplish my mission, I am going to focus on four aspects of the scientific/research market. I will spend a few moments trying to define what I call the scientific market, which I also will refer to as the research market. I will define the kinds of lasers that we see going into that market, and I will talk briefly about who the major players are that serve that market. I'll then cover some economic data regarding the market. I am also going to give you my forecast for 1988, and my concerns about 1988. Then I will move on to review some of the technology trends and key factors that I see driving this particular market, and I'll close with a brief summary in which I will try to highlight some of the things that I think we should all be watching as we participate in this market going forward.

This chart (Fig. 1) is a way of describing the scientific market. It is a complicated market. It involves applications in fundamental research. It involves applications in applied research. It involves the development of a variety of analytical tools. I have attempted here to describe some of the different applications that go on. I don't intend to get into any detail here. I think many of us know that the list of things that go on in the research market is almost endless, and we are surprised on virtually a weekly basis with what people are doing to do research with lasers. It is clear that research is done by a wide variety of disciplines: physicists, chemists, biologists, material scientists, people in medicine, people in the semi-conductor industry; again it is a wide range of targets. And we as an industry bring in a lot of different technologies to bear on this market. I believe it is safe to say that the scientific/research market is the most complicated market that any of us address. It is characterized by all of these different applications, and the various disciplines that we have to bring to bear on it.

As the backbone of our industry, this market essentially takes product from all of the lasers types that we collectively manufacture. This list should include the diode-pumped YAG products as well. It was left off the list. It is interesting to note that the list is, with the exception of diode-pumped YAGs, this list is pretty stable. These are the products that we have been supplying to this market for a long time. It doesn't say that there are some very interesting things going on within the various product lines, and I'll try to address some of

Figure 1

Scientific Lasers -- Main Applications

FUNDAMENTAL RESEARCH
- Frequency Resolved Spectroscopy
- Time Resolved Spectroscopy
- Scattering (Raman, Rayleigh, Brillouin, MIE)
- Non-Linear Spectroscopy

APPLIED RESEARCH
- Materials Research
- Light-Tissue Interaction
- Systems Development

ANALYTICAL TOOL
- Quantitative Spectroscopic Techniques

DISCIPLINES INCLUDE:
Physics, Chemistry, Biology, Medicine and Material Research

that as I talk about trends and factors in the market. But if you step back and look at the market, it is a complex market being served by a relatively fixed set of laser types.

This list was an attempt on my part to define some of the major suppliers in the business. I think it is clear to say that my company, Spectra Physics, and the combination of Coherent and their subsidiary, Lambda Physik, are the major suppliers to this market. We collectively supply well over half of the product that goes into this market. Other major suppliers are Lumonics, Cooper, Quantel, Quantronix, and Melles Griot. The list of suppliers to this market is pretty stable. With the exception of Lumonics, a Canadian based company, the manufacturers are all American. This is one of the market niches where American technology and American manufacturers do very well relative to suppliers from other parts of the world.

THE WORLD MARKET

Now I would like to move into a discussion of the economic outlook for the business. I will start by giving you my forecast for 1988. It is close to Moe Levitt's forecast that was brought together by Laser Focus. I'm projecting that the market will be up about 10% this year, with sales in the 138 million dollar range. The year-to-year growth that I am projecting is a little bit below what is in the Laser Focus estimate. There was a spurt of business in the fourth quarter of 1987, that I don't think was picked up in the numbers that you all pulled together. Business picked up very dramatically in both Japan and West Germany right at the end of the year as those two governments pumped their systems for entirely different reasons.

This market has grown at a fairly consistent rate in the past. If you go back way in time, the research market has grown in the 10% to

15% rate on average year after year. An estimate for 1988 of 10%
growth is really consistent with that. The bottom bullet though is the
one that is important for all of us to think about, and that is that
the growth in this market is very definitely driven by government
funding on a worldwide basis. As it has been pointed out already this
morning, I think there is a lot of nervousness about what the inten-
tions of the U.S. government will be in 1988 relative to our funding
across the board and particularly how that will affect the research
market. So I make a projection of 10% growth in 1988, but I am nervous
in making that. I can tell you that we at Spectra Physics are watching
for signs and indicators in the first part of 1988 as closely as we
know how.

This is some information about how this market breaks down worldwide (Fig. 2). In 1986 we felt that the domestic market accounted for
about 45% of the worldwide research market. And in 1988, due to growth
at rates about that of the United States in both Europe and the Pacific
rim, those ratios change. Europe increased to almost 45% of the total,
the Pacific rim increased to over 15%, and as a result the U.S. market
as a percent of the total market dropped to a number that is down in
the 40% range. Please keep in mind there is some error band in all
these numbers. I think it is important for everyone to be aware that,
at least from our perspective, there are some changes going on in where
the markets are on a worldwide basis. My prediction for 1988 is that
the U.S. will continue to grow at a slower rate than both Europe and
the Pacific rim market which is mainly the Japanese market. And at the
end of 1988, the U.S. share of this total worldwide research market
will be below 40%.

THE U.S. MARKET

In talking about the outlook for this particular market, I would
like now to turn to some data that I was able to pull together for the
U.S. market alone. The data that I will be covering in the next three
slides is information that came from Chemical and Engineering News. I
found it very difficult to get good, reliable data from major international country markets that could compare with what I was able to get
here. It just doesn't appear to be available in a form that, at least,
we have the ability to deal with.

So this is a look at where money came for research and development
in 1987 (Fig. 3a). This is a projection that was made by Chemical and
Engineering News. The summer of last year shows that the total
research and development funding for the U.S. market was estimated to
be 122 billion dollars. I would just like to point out a few things
about these numbers. First, the year-to-year growth of 6% projected
for 1987 is really modest growth. For the last decade R&D funding in
the U.S. has been growing at a rate more like 10% per year. I believe
that this is the lowest year-to-year number in the last decade.
Another thing to note is that the funding from industry sources is now
below that of the federal government. For a long time these two
sources were pretty much running neck-to-neck, and all of a sudden
industry is lagging. Within the federal government, about 70% of those
funds come out of the DOD. The remainder come from the space program
and a variety of sources directed to non-military applications.

RESEARCH MARKET
WORLDWIDE DISTRIBUTION

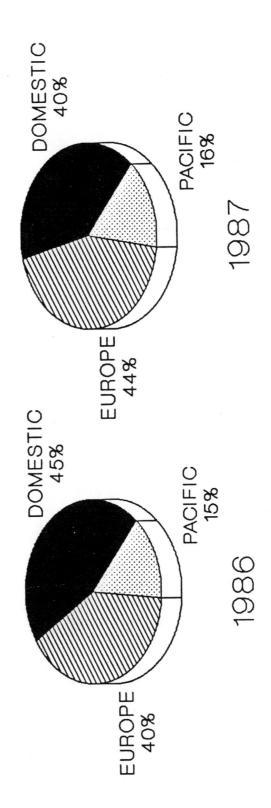

Figure 2

Figure 3a

Sources of R&D Funds (1987)

	($B)	Chg %
Industry	$ 58.1	5
Federal Government	60.0	7
Universities	2.7	8
Other	1.5	7
TOTAL	$122.3	6

Source: "Chemical and Engineering News"

Ⓢ Spectra-Physics

I expect in 1988, growth of funding in the U.S. market will be no greater, on a year-to-year basis, than it was in 1987. And depending on some decisions that get made by the U.S. government here in the first part of 1988, the year-to-year growth could be less than 6%. I think that what the U.S. government does is really key.

This was a projection of where those funds were going to be used. From a use standpoint, more than 65% of these funds, of the 122 billion dollars, go into what the government classifies as 'development', leaving something less than 35% for research applications. That's the pot that we are attempting to sell lasers into. So, in a gross sense, we are after funds that come out of the smaller of the two funding definitions. In 1987, I believe that the development funding grew faster than funding for basic and applied research. I think that trend will continue for the most part in 1988. You can see from this that industry consumed a very big part of the total amount of money -- 73 or 74% of the total. That has been a constant percentage for a number of years, and I don't think that is going to change in 1988. I think we are going to continue to look to the industrial market as the largest total potential market for research and development uses of money.

So, those are gross numbers. This (Fig. 3b) is a whole different set of data, so you can't tie it very easily to what I have just shown

Figure 3b

Domestic Government Budgets -- 1987

($M)	Basic	Chg %	Applied	Chg %
NIH	$2938	-6	$1368	-6
DOD	996	-	2636	11
DOE	1063	12	913	-15
NSF	1423	13	86	10
NASA	986	16	1397	25
TOTAL	$7406	2	$6100	2

Source: "Chemical and Engineering News"

Ⓢ Spectra-Physics

you. This is a representation of money available from these five
funding sources for basic and applied research in 1987. I think, if I
step back and look at these numbers, the thing that is of most concern
to me is that in aggregate, both basic and applied research were projected to increase only 2% in 1987 over 1986. If you look at the mix,
you can see year-to-year projections that are all over the map from
declines to a very big increase in the applied side of NASA funding. I
can't sort that out particularly well. My concern, though, in looking
at 1988 is that if 1987 only had funding growth of 2% in these two
categories, and if the government is going to consider government
funded research and development activities as one of the areas they are
going to look to in attempting to reduce the budget deficit, then the
increase in 1988 over 1987 most likely isn't even going to be 2%. And,
in fact, it could contract in 1988 and that has to have some kind of an
impact on the total accessible market for us in the laser research
area. That is why I am nervous in making a projection of 10% year-to-
year growth overall.

So, if I step back and try to summarize my outlook on the economy,
I am projecting 10% a year-to-year growth with nervousness. I think
the foreign markets are going to continue to grow as a percent of the
total. I believe that industry, as a user of funds, will continue to

exceed the government and that the uncertainty in all of these projections, the greatest uncertainty, is what the U.S. government is going to do to address the budget deficit and how that will impact their funding for research and development.

TECHNOLOGY TRENDS
Now I would like to move into a discussion of key technology factors and trends (Fig. 4). These are things that are directly impacting the laser market.
The first trend is a continuing trend. And in fact, a lot of what I am going to be talking about here is trying to provide you with some continuity in what we see going on in the research market. I'll try to

Figure 4

Technology Factors and Trends

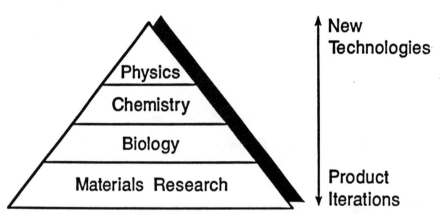

- Continued Demand for Ultimate Performance (Top)
- Increased Demand for Turn-Key Systems (Bottom)
- Growth Within Every Discipline

Ⓢ Spectra-Physics

help you understand what we are saying here, but this is a continuing trend toward products at the high end of the research market that are based on new technologies. Products up here at the top that serve mainly the physics and chemistry sectors within the research market. At at the same time, a continuing trend toward products in the lower sector of the research market that are based more on design interation.

When we talk about high end products, we are talking about products that have the following kinds of characteristics: There is a lot more customization involved; they tend to be more one of a kind they are more systems oriented; they are a lot more complex; our customers are very demanding in the performance they want; they (the customers) are very uncompromising; and supplying those products is a lot more risky for we, the laser manufacturers.

At the low end of the research market where a lot more product interation goes on, we see customers wanting more catalog oriented products. There is less systems integration demand on we, the manufacturers. And there is an ever increasing requirement for more and more reliability in the products.

Another part of this trend that I will point out is that throughout 1987, we saw serious price pressure on a worldwide basis in the products we were selling into both ends of this research market. I believe that price pressure is most likely going to continue in 1988.

The second factor that I would like to talk about is the ever increasingly complex design box that we laser manufacturers have to deal with (Fig. 5).

Figure 5

Technology Factors and Trends

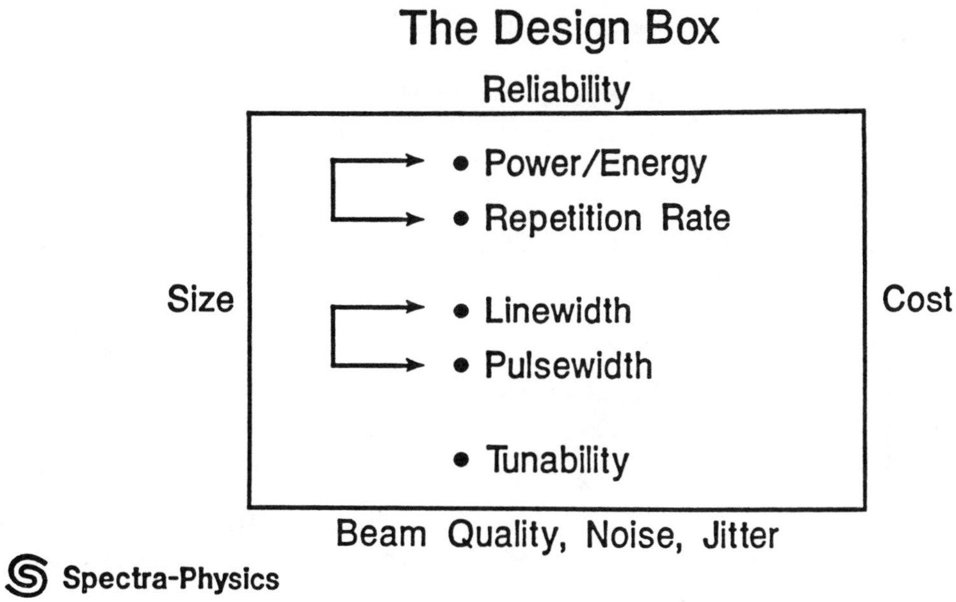

The Design Box

Reliability

Size

• Power/Energy
• Repetition Rate

• Linewidth
• Pulsewidth

• Tunability

Cost

Beam Quality, Noise, Jitter

Ⓢ Spectra-Physics

What I am trying to show here is that when we design a new laser with certain beam quality, noise, jitter, power levels, etc., we are really constrained by the size of the product, what physical size can go into the customer's lab, what price he will pay for that product, even if it is a top end research product, and what kind of reliability we have to design into the product. We are making trade-offs between power and energy versus what kind of rep rates we provide; linewidths versus pulsewidths kind of trade-offs, and we are dealing on an increasing basis with the need for more and more tunability. So it is a very complicated design arena that we are working in. I think the implications from that are the following for all of us as laser manufacturers: There is a dramatic increase in what it costs to develop a new product. Years ago we could develop a new laser for 250 thousand dollars. Today some of these lasers are costing 2 1/2 million dollars and more to develop. I think that is going to continue to go up. Because of that increasing cost, and because of the increasing complexity of what makes up the design box, I think the barrier to entry of new players into the research market is going up. One result of all of this is there may be a need for some technology-driven consolidation in the industry in order to obtain the critical mass needed to participate in this market and to deal with the increasing complexity that we see taking place in the design area.

The third trend is a continuation of a trend that we have noted in the past. That is a trend toward expanded coverage of the power and rep rates spectrums (Fig. 6).

Figure 6

Technology Trends and Factors

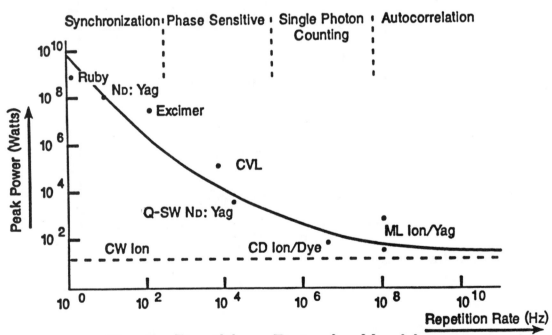

- Every Reptition Rate is Usable
- Almost Every Peak Power is Available

Ⓢ Spectra-Physics

This Figure shows where various types of lasers tend to fit into that arena. I think in today's market, it is important to note that there continues to be a need for basically every repetition rate, and there seems to be some kind of a need for just about level of peak power. It was pointed out earlier in the talks, there is activity across a broad spectrum in lasers going into our general markets, and in particular, into the research market. Certainly this is one way to think about it.

The last trend I would like to speak to is a continuation. But I think it is a continuation at a slowing pace toward shorter pulses on the one hand, and toward narrower linewidths on the other hand. In both cases, I think we are running up against some constraints. Achieving narrower linewidths is becoming an area where we are being limited by our ability to detect the linewidths. I think we are being limited by our ability to prepare reference standards. In shorter pulsewidths we are starting to run up against some of the pure limits of physics. But, the activity will continue, and we will see demand for shorter pulses, demand for narrower linewidths, and I think our industry will try to respond.

NEW LASER PRODUCTS

As I mentioned earlier, although the cast of laser types serving this market is relatively constant, a lot happened in this last year in terms of new products. I think the rate of introduction of new lasers for the research market accelerated, if anything, in 1987. And this is a list of some of the new introductions that caught my attention (Fig. 7). I know it is an incomplete list, but there were certain things that happened in 1987 that I thought were exceptional in the way of new products, and this is a list of some of them. They are not presented in any order of importance to the research market. At the top of the list diode-pumped YAG products became available from a number of manufacturers and in a number of forms including frequency doubled green, Q-switched and really high performance single frequency. Both Coherent and we (Spectra Physics) announced new levels of UV performance in ion lasers, which have expended the use of those products, as pointed out. A new type of CW YAG laser, the Antares, was announced by Coherent that involved a new cavity design, and more systems integration than was shown previously in those types of products. Spectra Diode Labs announced a 1 watt CW diode array, and that horserace just continues to go on, never ending. It is a little terrifying for the old gas laser guys to see those things coming along. Injection seeded and higher energy energy solid state YAG lasers were announced and delivered by both Spectra Physics and Quantel. New forms of excimer lasers were developed, announced and made available by Lambda Physik, Questek and Lumonics. And actually, the number of manufacturers of excimers expanded pretty dramatically with some new lasers being made available in Japan. And finally, at the bottom of the list, several new solid state materials were made available that are going to signal some very exciting new products in the future for the research market.

Future trends in products for the research applications: Shorter pulses and more energy per pulse will continue to be a driver of product development here. I think broader wavelength coverage with reliability improvements will continue to be a fact. I think you are going to see more systems integration by the manufacturers, and more of a

Figure 7

New Products in 1987

- Green, Q-Switched, and Single Frequency DPY

- New Levels of CW UV from Ion Lasers

- New Type CW YAG Laser Involving New Cavity and more System Integration

- 1 Watt CW Diode Array

- Injection Seeded and Higher Energy Solid State YAG Lasers

- New Forms of Excimers

- New Solid State Materials

Ⓢ Spectra-Physics

commitment by manufacturers to guaranteeing systems level performance. We are going to see new types of tunable lasers. We are going to see smaller footprints in our laser systems, mainly driven by solid state systems.

I would like to close by sharing some of my observations and beliefs about what we can expect in the future from this market.

First, I think the market for scientific lasers will continue to grow slowly. It is vulnerable in the U.S. to efforts by this government to control the budget deficit. And as I pointed out, we have to watch that carefully.

Second, a strong appetite exists among customers for new technologies that do new things. That is the never ending driver of this research market at the high end.

Third, manufacturers will need to iterate products over time to a greater extent than in the past in order to broaden the applications for the products and get a return on these high investments that we are making at the front end in the development of new core technologies.

Fourth, there is increasing competition among laser types for domination in individual niches. It used to be that the territory, the playground, was staked out for ion lasers, and it was different for solid state lasers, etc. That is no longer true. We are fighting among the technologies for domination and individual niches.

Fifth, customer service associated with these ever more complex products is becoming more important and more expensive. This industry of ours has to mature in this area in order to provide the levels of service that our customers demand, and learn how to make money at it. I don't think we have done a very good job of that.

And finally, given the relatively poor financial position of many manufacturers in this industry, and I am not talking necessarily about balance sheet strength or weakness, I am talking about market valuation, I think the potential exists for increasing consolidation in this industry. In fact, I think in the next five years we are going to see a significant consolidation among the suppliers of lasers in this particular niche and in our industry in general.

AUDIENCE QUESTIONS
Q: PLEASE DISCUSS THE NEWER TECHNOLOGIES ASSOCIATED WITH NON-LINEAR CRYSTALS.

A: I am talking about the availability of materials like ti-sapphire, beta barium borate, and other materials which hold the promise to help us achieve tunability in different ways, mixing with different levels of performance, frequency doubling, that gives us new levels of performance in the marketplace.

Q: COULD YOU EXPLAIN A BIT MORE THE CONCEPT OF INCREASED SYSTEMS INTEGRATION.

A: Two years ago if a customer at the high end of the research market wanted an amplified femtosecond system, he was probably willing to buy an amplifier from one manufacturer and a femtosecond system from another manufacturer. In fact, he might buy the pieces for that from more than one supplier, and go into the effort of integrating that system together and achieving whatever results he wanted. We see a trend in today's market for that customer wanting the manufacturer to guarantee a certain level of energy per pulse for 90 days or whatever the period of time is, so he is thrusting that burden back on the manufacturer. I believe our industry is going to have to respond to that one way or another.

BARCODE SCANNING MARKET
THE FASTEST GROWING MARKET FOR VISIBLE LASERS

Richard Bravman
VP of Marketing, Symbol Technologies

I will admit to the fact that the title of talk was suggested, however, I do believe it is quite apropos. When we take a look at some of the market developments that we are seeing now, I think you will agree that perhaps the term 'exploding' is not inappropriately applied to the market.

Let me just introduce myself again. My name is Rich Bravman, Vice President of Marketing for Symbol Technologies. Symbol is a manufacturer of barcode scanning equipment. We are based on Long Island in New York. And under that particular definition of our business, barcode scanning, we apparently enjoy the position of the world's leading manufacturer of that category of product.

What I will do today is try to give you some basic familiarity with what the barcode scanning market is all about, and what some of the implications of it are back into the business of laser technology. First, for those of you that are not familiar with barcodes, I think it is important to understand that the range of applications for barcode scanning today is considerably broader than what you may have experienced as a result of your day-to-day layman's life. I think most of us have come across the application of barcoding in retail point of sale application, but even within this oldest and most traditional of market application area, we find there have been significant developments very recently.

Supermarkets have been using scanning technology since the early 1970's timeframe. They were pretty much alone in that regard up until quite recently. It is only in the last two, three or four years that retail sectors beyond the food store have gotten into barcoding in a big way. Now that growth has become quite explosive -- again the term applies. We find today that mass merchandisers, drug stores, liquor stores, book stores, department stores, with very large requirements for large numbers of systems to support their needs are now coming on-line with barcode scanning.

Importantly the applications for barcoding within retail have now spread beyond point of sale. There are movements afoot to use barcoding to control the flow of merchandise all the way through the chain of supply leading up to the final point of sale. In one notable example there is a group looking to standardize the application of barcoding for everything involving the textile industry ... to the point that everyone that manufacturers yarn that is sold to textile manufacturers, that is sold to apparel manufacturers, that is sold to retailers, will all be under a common umbrella with a standard, defined application for barcoding. So there is a whole range of applications under this general umbrella of retail.

<u>Factory</u> <u>automation</u> has been a long-term user of the technology for applications such as work and process tracking, cycle counting and maintaining better control of their inventory.

The health care industry has a tremendous set of motivations to use the technology -- basically relating to the fact that they are looking not only to achieve productivity improvements over what might be possible via manual methods, but they are looking to deliver more effective health care. There are some very frightening numbers that people have cited recently: I'll mention one. An American Hospital Association study found that on average, when you are in the hospital you will receive one medication error per day. Across the entire United States, the average medication error rate is one medication error per patient per day. Barcoding has been identified as a very big part of the solution. They will be barcoding under standards that are currently being developed and introduced into the market, barcoding wrist bands, barcoding every item of medication that is to be administered to patients, basically barcoding everything that comes in contact with the patient. With the use of barcode scanning as an absolute 100% validation that the correct medical procedure is being accomplished at the correct time.

Office automation. Even in the general commercial markets, barcoding is being used for file folder tracking. One major insurance company found that they were 70% more productive in terms of clerical personnel in keeping track of files in a major claims processing center. Instead of losing the files constantly and having to go running around trying to find them, barcodes identifying the files were used to maintain a database showing at any time exactly where file folders were in their processing center.

The government has become a major application area for barcoding. Sometime ago, in 1982, the Department of Defense issued a military standard for the application of barcoding in military logistics applications - shipping, receiving, inventory taking, etc. Since that time, they have let prime contracts totaling to well over one hundred thousand barcode scanners to be delivered over the next five to ten year period. So they are a major consumer of the project itself.

In fact, this list could probably fill many slides if I showed it in full detail, and showed all the areas where there is growth potential today. That breadth of market represents one of the great strengths of our business that we have chosen for ourself. It also represents one of the true marketing challenges that we face in the sense of the need to develop mechanisms for reaching out to all of these various marketplaces, setting up various distribution channels, speaking the language of all these different user groups in terms of what their needs are. This is one of the real challenges we face as a company.

STANDARDS AND AUTOMATIC IDENTIFICATION

One of the major elements that has catalyzed the recent growth that has occurred in the market has been standards development activity. Various vertical industry groups, over the last three or four years, have put in place standards for how barcoding is to be applied in their industries. In each and every case, the availability of a standard has immediately preceded a rapid increase in the size of the market. People feel confident when they feel that they are able to implement a system which is standardized across their industry. Some of the specific examples I cite here include the UPC standard, the oldest going back to the early 1970's for retail; VICS, which stands for Volunteer Interindustry Communication Standard - that is the textile, apparel, retail standard that I mentioned a moment ago; The Health Industry Business Communications Council has a set of standards for barcoding in health care; Automotive Industry Action Group, similarly for the automobile industry and those people supplying them; LODMARS is the standard that the military adopted in 1982; and the telecommunications industry is now bringing forward a standard under the banner of Telecommunications Industry Forum. And this list could be twice as long if it was to be fully inclusive. What is an important trend, though, is the development of standards for the application of the barcoding technology.

It is important to understand that barcoding is a part of a broader technology umbrella called Automatic Identification. Automatic Identification simply put is a family of technologies designed to allow machinery to automatically identify the status, location or identity of an item. It is a very general definition, because, in fact, it is a very general set of capabilities that are provided. These are the technologies that range from barcoding through RF systems, voice recognition, OCR, machine vision, and mark sense, that fall under the general umbrella of Automatic Identification. We find today, however, that barcoding among these technologies has emerged as the dominant technology for general purpose application. And among the reasons why are the following:

Flexibility - the ability to work under many different application environments -- very harsh environments, various cost/price sensitivity levels at which systems are able to be delivered. In general it is a very cost effective system. Generally people implementing barcode based systems report return on investments or payback periods of less than a year.

Ease of Printing / Ease of Scanning. The ease of the ability of creating a barcode mark and the ease of scanning that barcode mark via a whole family of technologies, combined to allow barcoding to become as important as it is today.

A Very Robust System. The simplest thing that can be said is that it works. Other technologies tend to have niche applications where special characteristics set them up as a good solution for a particular niche application. But across the broad range of auto ID applications we see people pursuing today, barcoding has become the dominant answer.

LASER SCANNING

The prime applications for lasers within barcoding include laser printing, laser etching, and laser scanning. There are a variety of laser devices which are being utilized in those technology areas, as delivered in products. I will spend most of my time talking about laser scanning today in as much as that is the area that our company participates in more directly.

In developing laser scanning products, there are a whole list of critical laser characteristics that must be kept in mind and that the system designer uses as criteria in selecting one device versus another. They include the full list that I have shown here, and again perhaps, there may be a couple more in some applications.

Wavelength is important. I will talk more about wavelength in a moment.

    <u>Divergence profile.</u> The operation of laser scanning in barcode reading is keyed directly against its capability of offering non contact reading ability. The ability to read a barcode over an extended working range. That characteristic of the barcode reading system is directly affected by the divergence profile of the laser itself.

    <u>Power consumption.</u> In many cases our barcode scanning devices are used in battery powered configurations, and so low power consumption is a critical factor in many of those cases.

    <u>Lifetime.</u> These systems are used in very high duty cycle situations, a typical retail operator may have tens of thousands of scans per day. There are applications where you have continuous duty, 24-hours a day for an extended period. The lifetime is certainly important.

    <u>Environmental considerations.</u> These scanners are used in environments ranging from the benign to the extremely harsh. In the automobile assembly line environment, for instance in General Motors, the workers on the assembly line are affectionately known to their own management as "the gorillas".

    We received a number of hand held scanners back that were particularly bashed up. We went out to the field to investigate. It turns out the way they were bashed up was that the cable was literally ripping right out of the bottom of the handle. The reason that was the case is that the operators had the scanner attached to a 90- to 100-pound trolley assembly which ran back and forth on a motorized track, so they could move from one station to the next on the assembly line. There was a button that moved this entire trolley under motor from one station to the next. They found that it was too much work to walk the three or four feet to push the button, so they would drag this hundred pounds (with the motor working against the trolley) by the cable and the average lifetime was on the order of a day and a half as a result of that. So that is kind of typical of some of the tough environments that these scanners work in, and of course some portion of those tough environments, or some characteristics of those environments translate down into requirements for the laser devices themselves.

    Specifically the <u>temperature and humidity characteristics,</u> temperature being most important. Our scanning devices are often used in normal outdoor ambient conditions, so operations in Minnesota in the middle of the winter is a very real world system requirement in some cases.

    <u>Cost.</u> Looking at the numbers that we are talking about here in terms of the volume of equipment that people are buying, certainly the cost is a major factor in their ability to justify the application of the technology.

    <u>Ruggedness.</u> We've mentioned that.

    <u>Electrical characteristics, threshold current and voltage drop</u> have to do with their incorporation in a particular system's architecture.

THE EVOLUTION TO DIODE-LASER SCANNERS

    One of the major transitions that is happening in our particular application for lasers, just as it is elsewhere, is the overall move from helium neon tube based technology to laser diode based technology. I am speaking to you today just about at the midpoint of that transition range. The transition began perhaps about two years ago, a year and a half ago as a matter of fact.

Let's take a look specifically at a product family that characterizes this evolution. This is the helium neon based product that we have been manufacturing since 1982, it is the LS7000. We have something over 100,000 of these units installed in the field in a whole variety of applications. This was introduced in 1982 and it was the first fully practical hand-held laser scanner to be offered into the marketplace. In June of 1986 we introduced the LS8000, which is a laser diode based equivalent of the same basic functions. These two devices basically perform the same operation. Their performance characteristics vary somewhat, but from an overall point of view you can think of them as being apples and apples comparison.

In terms of non-performance issues though, there are significant differences. The LS8000 is half the weight of the LS7000. The power consumption is about 1/7 that of the LS7000. The amount of internal room within the scanning device that is taken up by the active circuitry has been cut down to the point that the entire handle has been freed up to have additional circuitry, additional functionality added to it. So, Jon Tomkins was talking about systems integration as a trend towards more systems integration, well indeed that's happening in our area as well. What we are finding now is that functionality that used to be in separate assemblies from the scanner, is now being built into the scanner as a result of the miniaturization possible through the move to laser diode technology.

So, the driving factors that motivate the move from helium neon to laser diode are summarized in the five points above:
1), power consumption 2), cost 3), ability to work under tougher environments 4), smaller size 5), and enhanced ruggedness. All of those are positives that one gets in going to the laser diodes.

The major trade-off though relates to the wavelength. All of you know of course that the wavelengths that have been commercially available in diode type lasers have until recently been in the range of 780 to 750 nanometers, and only now are starting to appear in prototype quantities at wavelengths in the range of 680 nanometers. The wavelength is important in barcode scanning applications.

First, differences in the way in which substrate material, the label material, or whatever the barcode symbol happens to be printed on or etched on -- the reflectance characteristic of the substrate varies as a function of wavelength. As one specific example, one of the most common ways to print barcodes today is via thermal printing, where thermally sensitive paper is used in a thermal printing device that heats up the paper to form the pattern of the barcode. It turns out that certain types of organically based thermal paper do not have the same, as a matter of fact they have radically different reflectance characteristics under 780 nanometer illumination as compared to 633 nanometers illumination. Specifically, that piece of paper that looks very white, highly reflective at 633 nanometers looks almost dead black at 780 nanometers. So the same symbol that can be read perfectly well with a helium neon device cannot be read with a 780 nanometer based laser diode.

A similar effect occurs in terms of reflective differences in ink used to print symbols, specifically color ink that is often used in retail applications where people try and blend the barcode symbol in with the packaging of the product. You see people using blue inks and green inks, etc., to print the barcode image itself. It turns out that certain shades of blue and green ink are just the opposite to the thermal paper situation. Instead of looking dark as they do to 633 nanometer illumination, all of a sudden they become light, so barcodes printed with blue ink on white background when viewed by 780 nanometer laser light look like a white on white symbol of insufficient contrast to pick up the return signal and accomplish the decoding function.

As a result these two characteristics of the symbol reflectance proscribe certain applications traditionally for the laser diode scanners. In situations where you have a fully open system, where you don't know which kinds of symbols you are going to be seeing, and so you may be seeing thermal paper, you may be seeing blue ink and green ink, then this type of scanner traditionally has not been able to be used. We will talk about the change as we go to 680nm in a moment. The one other factor that is not mentioned on this slide relates to the visibility, the human visibility. In a hand held device, of course, the ability to see where you are pointing, to see that the scanning beam is successfully crossing the pattern of bars and spaces that comprise the mark is an important characteristic of the device. 780 nanometer illumination, of course, is not directly visible to the human eye, at least at levels sufficient to provide the feedback that I just mentioned. So as a result we have to, in this system, incorporate a secondary aiming system. We use a conventional 630 nanometer LED, roughly aiming the diode to allow the operator to sight the symbol. That is a disadvantage in that the light from the LED is considerably less bright and less distinct than you get from the helium neon device, and so in bright ambient light conditions the aiming is not quite as user friendly. It also adds costs, some of the costs you take out by going to laser diode, you put back in by having to put in secondary aiming mechanisms, etc. Those are the main drawbacks in laser diode.

Now as we go to a visible laser diode, at 680 nanometers, these drawbacks will largely go away. As a matter of fact, all of the per-formance characteristic issues of substrate and ink reflectance go away by the time you are down to 680nm. It turns out that you pick up the ability to read blue ink and green ink down below about 750 nanometers, and you pick up the ability to read thermal organic paper below about 690 nanometers, so once you are at 680, which is where the current devices are being introduced, all of the functional characteristics are now fully equivalent with what can be expected from a helium neon based device. The one disadvantage remaining would be human visibility. It turns out that at 680nm the eye response is not quite up to picking up perceived brightness at the level that you would at 633nm, and so the apparent brightness to the operator is still one of the important design criteria that we are working very hard at in looking at ways of taking advantage of the new visible laser diodes.

MARKET SIZE

Let's just take a look very quickly at our projections for the size of the barcode scanning market (Fig. 1). Again, this is the

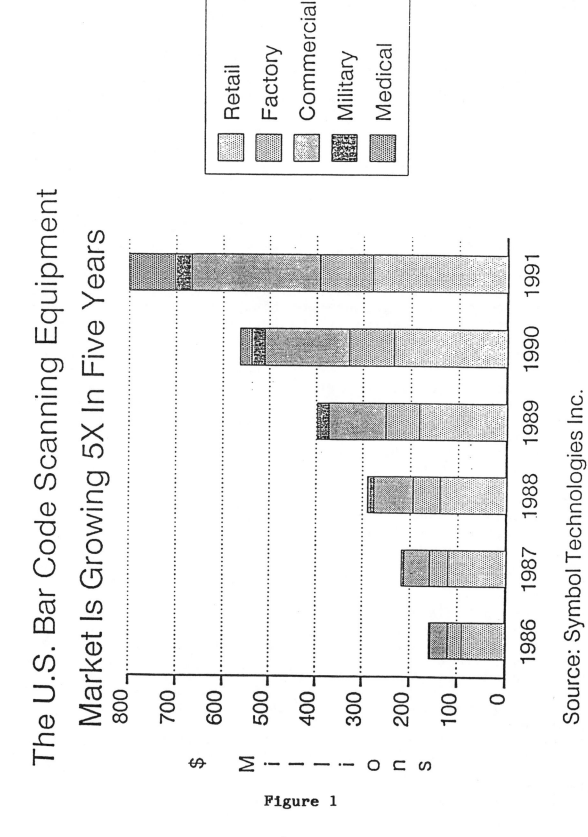

The U.S. Bar Code Scanning Equipment
Market Is Growing 5X In Five Years

Source: Symbol Technologies Inc.

Figure 1

systems market, not the direct dollar market in terms of units of component. You see that our projection calls for a sizable market to begin with. We believe that in 1987 the total U.S. barcode scanning market was in the range of 200 million dollars. We project that growth over the next 5 years through 1991 will result in a five-fold increase in the size of the overall market. You see that that growth is spread among growth which will occur in each one of the major market segment categories that we talked about a little bit earlier. Our experience over the recent years would indicate that this trend is realizable, and we very strongly believe that, in fact, this is the market growth rate that we will realize. The distribution between the different market segments may be somewhat different.

We believe that the particular category of product that our company is most strongly participating in, the hand held laser scanning market, will also account for a very big percentage of this total growth. As you see, in 1987 the delivered value of hand held laser scanning equipment exceeded about 55 million dollars by a bit. By 1991 we believe that will grow to something in excess of 300 million dollars. Today our company is the clear leader in this category, with a market share that is measured at something in excess of 90% in hand held laser scanners.

This information may be most useful to you as background reference. There is a trade association that represents all of the manufacturers participating in automatic identification, called AIM (Automatic Identification Manufacturers). And you see the range of programs and capabilities that the organization offers. So if you would like some further resource information about what is going on in barcoding, and would like to get access to detailed educational material, etc., this organization would be a useful one to know about. You may contact AIM at 1326 Freeport Rd., Pittsburgh, PA 15238.

AUDIENCE QUESTIONS
Q: WITH THE 680nm LASER DIODE, WILL REFERENCE LED'S STILL BE REQUIRED?

A: That is an open design issue at this point. Depending upon the ambient light environment that you are working in and depending on the optical path that your system design is based on, the answer may come down either yes or no. If you end up throwing away a fair percentage of the diode light in aperaturing it to deal with the divergence pro-file that I mentioned, then you may end up having to use supplementary aiming systems if you can figure a way of maintaining or saving as much of that available light energy as possible, delivering it to the symbol, then the answer may very well come back that it is not required. That is an open issue that we are wrestling with just now.

Q: IS THE LABEL REFLECTIVITY VARIANCE THE REASON WHY AT A POS CHECKOUT THEY MAY HAVE TO RUN THE LABEL SEVERAL TIMES?

A: Well, there are a variety of reasons why, in traditional type supermarket scanners, you may see them doing rescanning. One, symbol quality unfortunately is not fully uniform. As a matter of fact, the way our company got in business some years ago was designing a laser

based barcode verifier that would measure the quality of a printed
barcode symbol and tell you how well it was printed. Even though
barcoding is a very robust system, it can tolerate all sorts of printing distortions etc., it turns out that it is not 100% robust against
some of the carelessness that does occur. So, printed symbol quality
is one issue. The deck-type scanners, while they have this omni-direc-
tional read capability are projecting multiple lines, the scan pattern
density is not so deep that it is impossible to project a symbol
through the reading zone in such a way as to miss all of the lines, and
result in a no read. That is one of the advantages of hand held, is
that the eye/hand coordination zeros the active scanning area directly
onto the symbol. The first read rate is often higher with a hand held
device. As a matter of fact, Touche Ross, well known accounting firm,
just compared a study looking at slot scanning versus hand held, and
they found that the first pass read rate was significantly higher for
the hand held as compared with the highest performing slot.

Q: WHEN WILL PRODUCTS UTILIZING THE 680 nm LASER DIODES BE AVAILABLE?

A: Within the next twelve to eighteen months on the outside. There
are already prototype devices that are being shown. Our posture being
the leader in this particular product category, is that we will not
introduce a product based on immature component technology. Until the
manufacturing processes are under control, we don't feel that we can
expose ourselves and our customers to the possible interruption in
supply of the components or variability in key design characteristics
that right now is still the rule. I would say that 18 months is very
much an outside figure, 12 months is probably a more realistic outside
figure, and it would not be surprising to see something in this
calendar year.

Q: WHAT TYPE OF LED ARE YOU USING?

A: For the aiming system? To tell you the truth, I am not 100% sure.
It is one of the superbright LED's, the new generation of superbright
LED's. I don't know the specific device that we are using. It is a
Sharp device. One question that came up earlier that perhaps I could
speak to very quickly. In the figures shown in Laser Focus for the
size of the barcode market for laser diodes, there was a disparity
between the unit growth and the dollar growth. The dollar growth was
shown at something like a six times multiple going into next year as
opposed to the unit growth of some 50% increase. I believe that some
of that will be accounted for as a result of the move from IR to
visible laser diodes. The pricing that we are currently seeing from
the manufacturers shows about a factor of 5, to as much as a factor of
10 difference between the component cost of the visible laser diode as
compared to the IR laser diode. So for a period of time, the revenue
growth in the diode component technology will exceed that of the unit
growth by a fair measure. Whether the numbers there are exact and the
phasing is exact in terms of time, is somewhat more problematic, but I
believe that the basic trend that is shown there, which is this sort of
bubble where pricing will go up for a time, is as a result of the
visible laser diodes coming on into real application.

OPTICAL MEMORIES
LASER'S HIGH GROWTH BUSINESS APPLICATION

Ed Rothchild
Chairman, Rothchild Consultants

Let me do a very quick review of the four basic types of optical recording, because there is an awful lot of confusion, not only in the marketplace but even within the industry. In this, I will also address, what kinds of lasers are being used now and some of the quantities of units that are already out there.

Optical storage really started about twenty years ago, actually 25 years ago by efforts to put television programs or movies in analog format on read only video disc players. And although that market has not achieved the spectacular success that was earlier predicted for it, it is also far from dead and, in fact, in the last couple of years is showing quite a resurgence, particularly in Japan, particularly following the introduction of the combination players -- that will play both a compact disc and an 8" or 12" video disc, and now there are also 12 centimeter, 4.72" compact discs with video on them. Some of you know that recently Sony and others introduced 8 centimeter, 3" CD audio singles ... cheap and dirty. In the video disc player arena there are almost 2 million laser video disc players installed. Although that is far less than anybody would have hoped for, it has been good enough in sales that there are now over 10 manufacturers worldwide making these, and the production of discs is over one million a month, and the production of players is somewhere around 30 to 40 thousand per month. It is reasonably healthy. And there are some spinoffs of that technology that are coming to market. One interesting one is from SOCS that gives you, on a 12 centimeter disc, 20 minutes of motion video at 500 megabytes of digital data.

By far to date the most successful of the optical memory technologies has been, of course, the CD audio player. Over 150 companies have licensed these. The production of players, all of which use diode lasers, typically 780 nanometer, the same as the video disc players with some wavelengths up to 820-830, but the diode lasers for the CD players are typically 780nm. Production is running about seven or eight hundred thousand a month, and a million a month or more around Christmas time. At the moment there are about 16 million CD players in use worldwide. And this year there will be over 100 million discs pressed. So that is really brought some opportunities forward. Now, with the exception of Phillips, all other compact disc player manufacturers, at the moment, are Japanese. The technology is spreading but it will spread elsewhere -- in Asia we will see the Korean companies, Taiwan, Hong Kong, Singapore, etc., jumping in very soon in both the CD and video disc player arena. (Goldstar, Sansung, etc.).

One of the more interesting opportunities for the laser people is in the 488 nanometer argon lasers used in mastering of the discs. Typically 100 milliwatts is what is called for there. Later in the presentation I will show you which companies already have factories up, and which additional ones are in production. Your market target, of course, is the disc people as well as the drive people -- for the writable drive arena -- all of whom experimentally do their own mastering of their writable masters in-house.

CD ROM is the computer related information delivery technology for digital or still image or some amount of limited motion video on the disc, and about 13 manufacturers have announced product, and most are shipping now. This marketplace has been taking off slowly, and there are between 50 and 60 thousand of these players installed at the moment. We see something close to 100 thousand being shipped this year. The thing that will really kick that market off is the availability of broad horizontal market appeal software. During 1987 one such product did come out from Microsoft, the leading micro-computer software vendor, called Microsoft Bookshelf, which puts 10 very useful databases for writers -- dictionaries, statistical abstract, zip code directory, spellcheckers, thesaurus, the world almanac -- that sort of information, plus 10 different common word processing software programs on one read only disk accessible from PC's and Apples, etc. There will be much more such software coming out, but until it comes out it is very difficult to forecast what the real growth rate of this technology will be. So that is the bad news.

The good news is although the technology in CD ROM has not grown as fast as people would like, the number of titles of discs being pressed, and therefore the number of masters being made, is much higher than forecast. The number of replicas of those is lower, but you as laser people don't care about that. The good news is how many masters are being made, and those will be the expensive 488 nanometer lasers etching the masters. There are spinoffs of CD ROM technology which I will get into. There is also the generic, usually 5 1/4" optical read only memory, which is not yet on the market. That will await the standardization to make that a viable market -- higher performance than the CD ROM. I won't go into that in any greater detail.

There is one other major optical format on the market now, and that is the optical card. The size of a bank credit card. The most commonly discussed one is from Drexler Technology up in Mountain View, California; 26 companies around the world have licensed that. There are both read only and writable versions of the card, 2 or 4 megabytes on that one. A Canadian company, Optical Recording Corporation, last month publicly demonstrated their much higher capacity optical card, which will have between 100 and 200 megabytes on the same sized card. The basic difference is that ORC is using one micron spot size, Drexler is using 5 micron size. Drexler wants to keep the media and the laser system as cheap as possible to maximize the proliferation of the technology. And Drexler's interest is selling lots and lots of media, whereas ORC wants to sell media and hardware and systems. And the Japanese printing company, DiNippon Printing has a read only card out with a much lower capacity, about 64K bytes. Basically it is one or two school or college textbooks on a card. So in essence, it is a print on demand system.

CD ROM is attractive, speaking of printing, because you can get the equivalent of a quarter of a million pages of computer output, word processing pages, stamped onto a read only disk in about 15 seconds, and mail them anywhere in this country for 22 cents. You would need a pretty sizable vehicle to move a quarter of a million pages of paper, and it wouldn't cost you 22 cents! So that is part of the excitement of read only.

WRITABLE OPTICAL TECHNOLOGY

I will focus primarily on writable optical storage. I will just touch on analog writable, which is analog video recorders, and yes, digital video recorders are coming. They have been demonstrated publicly during the last year. This technology at the moment is not a consumer product. The cheap machines here are running around $15,000. The most sophisticated ones from Optical Disk Corporation are running about $200,000, for the professional market (that is not including batteries). It will be in the 1990's frankly before we have either consumer recordable video discs or consumer recordable compact discs on the market. The 1990's will really mark the explosive growth in this technology, because at the moment it is really in the low-cost read-only mass consumer market with lots of diode lasers being shipped but the prices are frequently at or below cost. No one is making a lot of money in that. The recordable music and recordable video machines will become popular when the cost of the blank media becomes cheap enough.

That means that for a CD audio disc, it better not sell retail for more than $5 or $10 in the stores or people would have no incentive to give up tape, even with the rapid random access capability of disc. For video discs, it not only has to be cheap but you have almost the same recording length of time possible as you do with VCR's today. Currently one hour per side is the maximum. But with the advent of the shorter wavelength visible lasers we have heard about, and more powerful channel codes, and very interesting technology coming up in the 1990's - three dimensional optical recording, where at any given spatial location rather than having just X-Y coordinates, you could record various levels deep. By using sophisticated optical filtering or spatial or spectral hole burning, which is a rather "researchy" topic at the moment, you can get a thousand times as much data on the same area of disc surface, or card or tape. There is a rather technical "oh, by the way" that I didn't mention. The spectral hole burning technology at the moment works at cryogenic temperatures only and there have been great advances on that recently and no longer is it a chilling 4 degrees Kelvin the stuff works at. We have heard reports of up to a balmy 23 degrees Kelvin.

In the writable digital area, about 54 to 56 companies have either announced or demonstrated or are developing writable digital optical disk drives. That is the major focus of our business. This will be the major opportunity for high capacity type power, multiple diode, lasers coming in the future. There is also writable optical card drives out, and more coming. And, optical tape drives - very huge capacity. The Canadian company, Creo, funded by Canadian government money, will later this year or early next deliver four prototype machines, at about a quarter million dollars each. But on the same size reel that holds 9-track mag computer tape, loading it with optical tape instead of magnetic tape. Instead of holding typically 180 megabytes, it will hold 1 million megabytes or 1 terabyte, and have average access anywhere in that 1 terabyte in 22 seconds, and worst case of 1 minute. That is mind boggling. That is about 5,000 optical disks worth of information. It is not going to be a mass market obviously, but for the really huge applications in digital imagery such as satellite reconnaissance, seismic data collection, medical x-rays where you have orders of magnitude more information requirement that an 8 1/2" x 11" piece of paper to store, various levels of gray scale optical tape could be a very interesting alternative.

Very briefly, these are the main advantages and features of optical recording (Fig. 1). You can read them yourself if you have fairly large magnifying glasses around your necks, and I'll just comment on one point. Up to this moment all optical media we've talked about to date are removable. But that will not always be the case. And that will become an interesting opportunity for you. All optical media at the moment are removable because comparing the optical head distance to the media versus a high density magnetic head, optical heads are 10,000 times farther away from the disk than a mag head, so removeability is no problem. However, if you want to get really high performance, very high RPM, very fast access time, tighten up all the tolerances, and eliminate some of the problems of dust and scratches (optical media is rugged but not totally immune from dust and scratches as was earlier said). There is a very interesting case to be made for multiple platter, multiple spindle optical disk systems and subsystems just as we have with the very high capacity Winchester drives.

The pressure to develop this is coming from the big IBM and other mainframe shops who have literally acres of 3380 disk drives. And especially if you are talking about Manhattan or downtown Chicago, or LA or God forbid, Tokyo, where real estate is now 100 times more expensive. Just the savings of the square foot footprint of the equipment is enough of an incentive. An example: I was in to see Pfizer Pharmaceuticals in New York at their world headquarters recently on 42nd Street. They have 100,000 sq. ft. of raised computer floor at an annual burden cost of $160/ft. That means before they spend a penny on any equipment, they have a 16 million dollar justification for optical storage. If they can reduce tremendously the square footage devoted to mass storage, which you can do with optical storage versus magnetic, that is a tremendous incentive. So the mainframe companies and the super-mini companies are getting very active in this arena where you will have multiple platters and eventually read/write heads on both sides of disk, the same kind of comb arrangement you have with magnetic storage, but for the multi-platter assemblies these are going to be higher powered diode lasers than the simpler plain vanilla kind of optical drives that are on the market today. And yes, all the mainframe and mini-computer American and European companies, are interested in that, and a number such as IBM, DEC, Data General, HP, Siemens in Europe, are developing such products. The main focus of the Japanese companies has been on removable single disk arrays for the mass desktop computer market. This is not to say that the NEC's, the Hitachi's, the Fujitsu's, and eventually the Sony's and Toshiba's will not get interested as well in multi-platter optical disks, but their main focus at the moment clearly is lower performance, lower cost mass market stuff.

Figure 1

Features of Optical Memory Technology

* CAN MIX DIGITAL AND ANALOG DATA AND INDEX ON SAME MEDIUM

* CAN MIX DIGITAL DATA, STILL IMAGES, MOTION VIDEO, AUDIO

* MUCH HIGHER STORAGE DENSITY THAN MAGNETIC OR FILM MEDIA

* MUCH LESS COMPUTER ROOM SPACE

* REMOVABLE MEDIA

* HUGE ONLINE JUKEBOX LIBRARIES ACCESSED IN 5-20 SEC.

* NO BACKUP OR RESPOOLING NECESSARY

* MUCH LONGER ARCHIVAL LIFE THAN MAGNETIC MEDIA

* VERY LOW COST PER STORED MEGABIT

* RAPID RANDOM ACCESS

* LOW COST REPLICATION

* RUGGED, NO MEDIA WEAR

* READ-ONLY, WRITE-ONCE, & RE-WRITEABLE DRIVES AND MEDIA

Source: Rothchild Consultants, San Francisco, CA - 1 OCT 1987

WRITABLE DISC COMPANIES

To give you some numbers on the writable disk area, as I did on the read only .. at the moment there are somewhat over 40,000, actually about 45,000 writable optical disk drives in use worldwide. And we are estimating somewhere close to 100,000 more will be shipped this year. The real take-off will be in '89 and '90 when widely available numbers of magneto-optical reversible, or rewritable drives will be on the market. We will see a number of introductions this year. There were a number of announcements last year - Sony, Sharp, Olympus, all announced and demonstrated publicly 5 1/4" magneto-optical erasable drives, and everybody will be doing that this year.

Here is the list (Fig. 2) of companies developing optical disk drives, and it includes many of the major computer companies in the world, plus some companies that perhaps are not familiar to you. I won't go into details as to who these folks are in this presentation.

Optical cards. There are over twenty companies doing that.

Figure 2

<u>Writeable</u> <u>Digital</u> <u>Optical</u> <u>Disk</u> <u>Drive</u> <u>Developers</u>

ATG GIGADISC
CANON
CHEROKEE DATA SYSTEMS
C. ITOH ELECTRONICS
CONNER PERIPHERALS
DATA GENERAL
DIGITAL EQUIPMENT CORP.
EASTMAN KODAK
FUJI PHOTO FILM
FUJITSU
GENISCO
HEWLETT PACKARD
HITACHI
HONEYWELL SPERRY
IBM
INFORMATION STORAGE INC.
IOMEGA
JVC
KAWATETSU INSTRUMENTS
KYOCERA
LASERDRIVE LTD.
LASER MAGNETIC STORAGE
LOCKHEED ELECTRONICS
MATSUSHITA ELECTRIC
MAXIMUM STORAGE INC.
MAXTOR
MILTOPE
MITSUBISHI ELECTRIC

MOUNTAIN OPTECH
NAKAMICHI
NEC
NIPPON COLUMBIA (DENON)
NIPPON KOGAKU (NIKON)
NIPPON TELEGRAPH & TELEPHONE
OKI ELECTRIC
OLIVETTI
OLYMPUS
OPTIMEM
OPTOTECH
PIONEER
RCA
RICOH
SANSUI
SANYO
SCHLUMBERGER
SEAGATE
SEIKO EPSON
SEMI-TECH MICROELECTRONICS
SHARP
SIEMENS
SOCS RESEARCH
SONY
SUNDSTRAND DATA CONTROL
TEAC
TOSHIBA

Source: Rothchild Consultants, San Francisco, CA - 19 JAN 1988

In optical media, if we eliminate discussion of the universities and just talk about commercial organizations, there are 80 companies around the world, either actively building optical media now or expecting to get into manufacturing within the next two years. This list includes virtually all of the world's major magnetic media manufacturers, film manufacturers, and about half the companies if you include film and plastics as being in the broader area, chemistry. Half the companies in the world in the technology or planning to get into it, are chemical companies, which is an important point. Up to now, the last 30 or 40 years, most information storage has been dominated by the physicists. Now the chemists are getting into the game with the potential of much lower cost media, where the chemical companies can do more than just provide the polyester substrate but they can provide the substrate, recording material, the protective layers and even the cartridge the disks live in. That indicates the seriousness with which optical media is being taken.

Now, is there need for 80 companies making media when in the magnetic media drive business, which is a 30 billion dollar a year business, there are only about 20 companies making media? The rest are all private label deals. Well, clearly not, and there will be lots of mergers and acquisitions, and companies throwing in the towel. But it indicates the seriousness with which this being taken.

I said magnetic storage is a 30 billion dollar a year business, if we talk about end user systems revenues, not just the optical drive and media and laser component. In 1987 the optical storage industry hit 1 billion dollars worldwide. However, only about 10 to 15% of that total is the actual optical drive and media portion. Most of the rest, as in computer systems, is software, high resolution displays, the laser scanners and printers, and of course the networks, and a modest profit for the systems companies.

There are 13 companies making CD ROM, compact disc read only memory hardware. At the moment there are 50, plus or minus 1 or 2, companies around the world actually pressing CD audio discs. There is a market for mastering equipment there. There are about 40 more that are planned or are under construction. However, a problem has arisen. Up until last year there was about a 30% shortfall in supply of compact disc versus demand. The world production for discs caught up during 1987, but we finished 1987 with an installed capacity potential of 2 times the demand. I am anxiously awaiting the price drops in stores. We know the factory price of the discs dropped 40% last year, but I am still paying $16 a disc from Tower Records and I want to know why. Actually we know what the reason is, and that is that the record companies are building up warchests to stave off the inevitable assault of RDAT, Recordable Digital Audio Tape. (You could ask the gentleman from Sony who is in the room at lunch, why they launched that competition with the CD, I would be very interested). There are about 40 plants under construction around the world including in the East Bloc.

THE WRITABLE DISC MARKET

In the writable part of the technology, when we do our market forecast, (and I'll share it with you in a moment, our market forecast and where this technology will be in units and revenue at the end of 1990), we segment the market according to the size of the disk and the disk drive, not the capacity of it because that is almost meaningless given the tremendous developments in shorter wavelength lasers. As long as four years ago when I was in a 3M lab seeing 5 1/4" magneto-optic media with a raw capacity of 2 gigabytes (on each side of a 5 1/4" disk, 1 1/2 gigabytes net user capacity). Yet today, there are still 12" disks on the market with under 1 gigabyte, so the only meaningful thing to talk about we think is the size of the disk, not the capacity because the capacities with optical are doubling about every two years as they are with magnetic.

Fourteen inch has very limited market units but high value. At the moment there are only two vendors, Kodak for the commercial market (they get into real production later this year), and RCA for custom government multi-million dollar systems.

Twelve inch, until 1987, was the dominant form factor in optical storage. It still is in revenue, and will remain so through 1990, but in units the 5 1/4"'s, which got into more or less mass production during '87, surpassed them in sales of units for the first time. This will remain the most profitable area in optical storage.

Typically we are dealing with diode lasers, the most commonly used now are 40 milliwatts. Of course what everybody in the optical storage business wants is blue lasers yesterday and they want them cheap, and they want them powerful. We are very excited at this SPIE conference (the session Gary Forrest chaired yesterday morning), to have heard about a number of visible wavelength diode lasers from a variety of companies: Sony, NEC, Amoco Laser and IBM, and others, coming very soon. That will really lead to the high performance products we are talking about here.

Eight inch, very few of them have been sold. I will make the forecast - unless there is a spectacular increase this year (which I don't really expect), 8" will be dropped. 5 1/4" is the clear winner in smaller than 12" disk format. 5 1/4" will represent well over 90% of the units, but it won't represent anywhere even close to 30% of the revenue for the technology. That will be the 12" disk systems and up, much higher price, much higher performance. Eventually there will be a segmentation of the massive 5 1/4" market, and it will split along price and geographic lines. Basically the mass market low priced, plain vanilla ones if you wish, will be made in Japan and other Pacific rim countries. We are already starting to see Hong Kong, Taiwan, Singapore, and Korea get into the act. The American/European com-panies, although many are in removable single disk drives now, will be getting into the multi-platter, multi-spindle high performance, high price systems, where there will be a premium payable for high powered short wavelength lasers to really get the packing density up.

3 1/2" in the 1990's will pass 5 1/4" in units. Generally speaking, companies who are doing 5 1/4" drives will not introduce 3 1/2" drives for a least one to two years after the 5 1/4"'s come out. Corollary to that, is any 3 1/2" drives you see introduced in '88 will come from companies not doing 5 1/4", such as Verbatim owned by Kodak now, Seiko-Epson, Nakamichi, etc.

Will IBM be doing 3 1/2"? Yes. 5 1/4"? Yes. 8"? Yes. When? Real soon now.

At the end of 1990, this is a very busy chart (Fig. 3a, 3b) I realize, here is what we estimate what will be sales in 1990 and installed base. I segregate these into write once versus erasable or rewritable. 14" will be small in units; there only will cumulatively be about 9000 of these drives out, and then we have the revenues here for 1990, about 97 million dollars for the drives, and about 8 million for revenue and about 40 thousand cartridges. The big unit, the big, big revenue generator, will be the 12" write once drives, which will be used primarily for massive document storage, archival imaging applica-tions which take up so much more space than computer data storage. In fact, the vast majority of optical disk drives installed today are used in office document or engineering drawing, or medical imaging systems. When the rewritable or erasable drives really get into production, this year and next, that will switch and it will be mostly desktop computer

Figure 3a

Worldwide 1990 OEM Optical Drive and Media Market Forecasts

	WRITE-ONCE	RE-WRITEABLE	TOTAL
5¼" DRIVE SALES, K units	300.00	600.00[#]	900.00
5¼" DRIVE AVE. OEM PRICE, $K *	0.50	0.74[#]	
5¼" DRIVE OEM REVENUE, $M	150.00	443.10[#]	596.40
5¼" DRIVE CUM. SHIPMENTS, K units	628.50	876.00[#]	1504.50
5¼" DISK CART. SALES, K units	3555.20	5760.00	9315.20
5¼" 2-SIDED DISK CARTRIDGE PRICE, $	30.00	35.00	
5¼" MEDIA OEM REVENUE, $M	106.66	201.60	308.26

****

	WRITE-ONCE	RE-WRITEABLE	TOTAL
3½" DRIVE SALES, K units	0.00	300.00	300.00
3½" DRIVE AVE. OEM PRICE, $K *	0.00	0.40	
3½" DRIVE OEM REVENUE, $M	0.00	120.00	120.00
3½" DRIVE CUM. SHIPMENTS, K units	0.00	415.50	415.50
3½" DISK CART. SALES, K units	0.00	2655.00	2655.00
3½" 2-SIDED DISK CARTRIDGE PRICE, $	0.00	25.00	
3½" OEM DISK REVENUE, $M	0.00	66.37	66.38

****

	WRITE-ONCE	RE-WRITEABLE	TOTAL
TOTAL 1990 DRIVE SALES, K units	575.00	911.10	1486.10
TOTAL 1990 DRIVE OEM REVENUE, $M	1395.00	592.40	1987.40
TOTAL 1990 DRIVE CUMULATIVE SHIPMENTS, K units	1085.20	1303.91	2389.11

****

	WRITE-ONCE	RE-WRITEABLE	TOTAL
TOTAL 1990 2-SIDED DISK CARTRIDGE SALES, M units	5.47	8.48	13.95
TOTAL 1990 2-SIDED DISK CARTRIDGE OEM REVENUES, $M	285.76	273.21	558.97

****

	WRITE-ONCE	RE-WRITEABLE	TOTAL
TOTAL 1990 DRIVE AND MEDIA MARKET, $M	1680.76	865.61	2546.37

* Price per drive including controller, in 1987 U.S. Dollars.
[#] Includes both re-writeable only and multifunction drives.

Figure 3b

Worldwide 1990 EOM Optical Drive and Media Market Forecasts

	WRITE-ONCE	RE-WRITEABLE	TOTAL
14" DRIVE SALES, K units	5.00	0.10	5.10
14" DRIVE AVE. OEM PRICE, $K *	19.00	23.00	
14" DRIVE OEM REVENUE, $M	95.00	2.30	97.30
14" DRIVE CUM. SHIPMENTS, K units	9.10	0.11	9.21
14" DISK CART. SALES, K units	39.60	0.60	40.20
14" 2-SIDED DISK CARTRIDGE PRICE, $	200.00	600.00	
14" MEDIA OEM REVENUE, $M	7.92	0.36	8.28

****

	WRITE-ONCE	RE-WRITEABLE	TOTAL
12" DRIVE SALES, K units	220.00	1.00	221.00
12" DRIVE AVE. OEM PRICE, $K *	5.00	7.00	
12" DRIVE OEM REVENUE, $M	1100.00	7.00	1107.00
12" DRIVE CUM. SHIPMENTS, K units	368.00	1.20	369.20
12" DISK CART. SALES, K units	1548.00	7.00	1555.00
12" 2-SIDED DISK CARTRIDGE PRICE, $	100.00	130.00	
12" MEDIA OEM REVENUE, $M	154.80	0.91	155.71

****

	WRITE-ONCE	RE-WRITEABLE	TOTAL
8" DRIVE SALES, K units	50.00	10.00	60.00
8" DRIVE AVE. OEM PRICE, $K *	1.00	2.00	
8" DRIVE OEM REVENUE, $M	50.00	20.00	70.00
8" DRIVE CUM. SHIPMENTS, K units	79.60	11.10	90.70
8" DISK CART. SALES, K units	327.60	61.00	388.60
8" 2-SIDED DISK CARTRIDGE PRICE, $	50.00	65.00	
8" MEDIA OEM REVENUE, $M	16.38	3.96	20.35

****

* Drive prices in constant 1987 U.S. dollars include controller.

Source: Rothchild Consultants, San Francisco, CA - 1 SEP 1987

applications. But the biggest revenue chunk will be in the 12" drives, and they will pay a premium for high-performance lasers.

8" - sad sales unfortunately. I think it will go away.

5 1/4" - The big volume, but mostly low priced diode lasers will be the 5 1/4" area. At the end of 1990 we think there will be 1 1/2 million 5 1/4" drives out there. This is almost an order of magnitude lower than our forecast for a few years ago. Privately over lunch I will chat with people who want our assessment of what happened and why it is taking longer. Are we still bullish on the technology? Absolutely! But the ramp up, the hockey stick curve, was simply pushed back, but there are too many big companies in this for optical storage not to eventually succeed.

3 1/2". Although it looks much lower in sales than 5 1/4", it is starting one or two years later, and that will pass 5 1/4" in units in the 1990's. What this means is that adding 5 1/4" and 3 1/2" together, if we take for example, Dataquest's numbers on the computer business, which are the ones that we think are the most accurate, at the end of 1990 they said there would be almost 25 million desktop computers in use worldwide. Plugging our numbers in, we think 1 out of 14 of those computers will have an optical disk drive on it at the end of 1990, a 7.6% market penetration. So those are very conservative numbers. Many of the manufacturers now have much higher market forecasts than we do. Given the fact that we have been burned so badly in the past by optimistic forecasts, we have gone the other way.

Those are our forecasts for the writable drive area in read only CD ROM. Although there were more CD ROM units shipped in 1987 than writable drives, that will not be the case in 1990 unless there is spectacularly compelling software available. And here are our forecasts in 1990 that there will be still under a million of those units out. I have seen forecasts as high as 35 million units out, but that depends upon two huge markets, which are viable but we don't think will happen before 1990. And that is CD ROM units built into cars for mapping and navigation systems. The very first one appeared 6 months ago in Toyota's in Japan, but it won't be a big market worldwide before 1990. And the last huge market for CD ROM will be dedicated CD ROM machines built into telephones for yellow page and white page lookup, with advertising, with some still motion video, with literally the sound of the steak sizzling when you dial up restaurant ads.

AUDIENCE QUESTIONS

Q: DO YOU HAVE AN ESTIMATE OF THE TOTAL GROWTH WORLD-WIDE FOR R&D SPENDING ON OPTICAL STORAGE.

A: Yes. Now this includes video disk. In fact the majority of the numbers I am going to give you was for analog video disk development. Which basically meant that CD audio and writable were developed based primarily on earlier money spent on analog video disks. Two and one-half billion dollars to date worldwide is a rough estimate (cumulative) of what has been spent. The first area to go profitable was the CD area, although now with competition there is not much profit left in either the discs or the players. But that was the first one to go profitable because of the volume.

Q: WE ARE ALL EXCITED OVER THOSE NUMBERS OR DIODE LASERS. ARE THERE
     ANY NON-DIODE LASER MANUFACTURERS IN ANY SIGNIFICANT AREA OF THIS
     MARKET?

A: Only the 488 argon lasers for mastering, because the number of
plants doing the read only or the writable disks will be something
like 150 to 200 by the end of the decade and there will be say 50
companies making writable drives, every one of which is a candidate to
have one or more in-house mastering machines using fairly pricey 488
nanometer argon lasers. All the rest, with only a handful of exceptions, literally less than you could count on your fingers, will be
diode laser devices. The only non-diode laser devices on the market at
the moment are in the custom system RCA sells to the government, and
they have only ever delivered two. So that is not a volume market.
But lasers for mastering machines would be the exception to the diode
laser dominance.

SPECIAL PRESENTATION:
INNOVATIVE SCIENCE AND TECHNOLOGY PROGRAMS
THE RELATIONSHIP OF GOVERNMENT-SPONSORED PROGRAMS
TO THE LASER INDUSTRY AND MARKETPLACE

James Ionson
Director of Innovative Science & Technology
U.S. Department of Defense

The Office of Innovative Science and Technology is involved with the Strategic Defense Initiative Organization. This office, the Innovative Science and Technology office, is different from most research outfits within the Department of Defense, and quite frankly within the government. What we do is attempt, and I think we do more than attempt, we actually do it, we integrate the basic science engineering into a product. We don't separate out basic research, engineering research, product development, fabrication ... we do it all together. We have an end goal in mind and we bring whatever basic engineering sciences that have to be brought to bear on that particular mission to come up with that product. Now the product that we come up with obviously fits the Department of Defense mission, and specifically the Strategic Defense Initiative Organization missions. Even so, those products have cut across many lines of application ranging from medical applications, all forms of commercial applications, you name it and it is there. If you look hard enough, and that is the problem ... knowing where to look.

We do not carry the product out to the point where you begin fabricating it, manufacturing it and mass producing it. It is a bread board. A good example might be a sensor system, since it is a hot topic now, with a super conducting infrared detector system. And I will get to something that is more germane to your areas of interest in a moment, specifically lasers.

We have a budget which is relatively small compared to the budget of the SDI. It is about 5% of the overall budget, but 5% of three to four billion dollars a year is better than a stick in the eye. We have a great deal of flexibility with that money subject to the checks and balances of government procurement, which protect all of us. The way to get ahold of that money, and I am sure that is foremost in everyone's mind, is simply to send in proposals. Not big long multi hundreds of thousands of dollars worth of proposals, (in other words the amount of effort that you have to put into it), but white papers. These white papers basically give us a feeling for the type of idea that you might have, and we work very hard to find a place for it within our particular mission, our particular market area. It might be basic research, it might be engineering, it might actually be involved with brass boarding or bread boarding a particular product.

This brochure, this list of agents, that you see ... those are the points of contact. They are the experts in the field. We are coming out with another brochure (I don't want to sound like an advertiser here), but this is part of the information process. We are coming out with a new one, it will list all of the agents, all of the points of contact that you should contact in order to get into this program. The key here is not so much the amount of dollars because the amount of dollars, although as I said it is better than a stick in the eye, is relatively small. It is the leverage. It is the ability to help leverage a future marketplace, both within the Department of Defense and well as in the commercial sector.

LOOKING FOR SYSTEMS

So, this office is a mean, lean product development machine, if you will, for the Department of Defense. Let me zero in the scope, or focus in a little bit on something that is more germane to you, and that is electro optical systems. You notice I said, electro optical systems, not just lasers. Because, what we are looking for in the spirit of this office, is the system. That is our ultimate goal, not a component so much because in many cases when you develop a component; take for example an optical storage medium or system using trapped electron states, you name it, but one part of the system ... it is like having a body with no brain in some cases. It is very difficult to sell that. It is very difficult in many cases to find the marketplace.

There is a good example of a small company which we have funded, they have developed a very interesting optical storage medium using trapped electron states, and I can't go into the details of it because a lot of it is proprietary. I can tell you the name - Quantex is the name of the company. They may want to tell you about it, but they have asked me not to tell you and I have to honor that, it is their trade secret.

What they are missing is an efficient way to write on that medium and read out. To write onto that medium they need a very efficient visible solid state laser, and I can go into some details. They have no expertise at all in lasers. In order to read out, they need a detector, a quantum limited detector. They have no expertise in quantum limited detectors. So what we try and do is match them up with the appropriate individual within academia, small industry, large industry, in order to make this whole system come to fruition. That really is our business. It is systems in general.

Now, what is the government market? What types of systems are the government interested in, and specifically, the Department of Defense. Well, I can name a whole bunch ... there are imaging systems, space borne, air borne, under water borne, imaging systems in the active sense, you just don't stand back and look, you probe! How do you probe? You probe with a coherent source of energy - laser. But simply having the laser is not enough, it has to interface and couple with the detector system as well as the various algorithms that make some sense of this information that you are gleaning from whatever object you are probing. Communications systems that require sources, detectors, and the ability to, once again, assimilate that information. Designators, tracking systems, fiber optics - all of the associated materials which one would need to transmit and carry this radiation.

Electro-optical materials - and that could range from counter measure materials which would be necessary to counter laser attacks on optical systems as well as to protect in the commercial world optical systems from various spurious laser radiation. One can go into micro-electronics as x-ray lithography, maybe someday gamma ray lithography, we are trying to develop the gamma ray laser which is, at this stage, very difficult and we're not there yet. Optical storage media, optical computing.

The key word, however, is information. Collecting it, assimilating it, transmitting it, storing it. It is information. And information, in my judgment involves a market that is probably the most rapidly evolving market in the entire world. The information systems that you see are antique a year later in many cases. What does that mean? In my judgment what that means is that small companies, in fact groups of small companies, are ideally situated to work very fast, they are very flexible, and they can integrate much faster than some of the larger corporations. That is what we are looking for. It is very difficult for a large corporation to, which quite frankly in many cases are set in their ways, to accommodate a rapidly evolving marketplace. And the information/electro-optic marketplace is incredibly rapidly evolving.

RELEVANT LASER TECHNOLOGIES

What are the associated technologies? And specifically, laser technology that could have an impact in this marketplace? There is a whole potpourri of them, and let me just mention a couple. This is research that is coming out of the academic world, coming out of the small business world, and if you just look at the piece of research in and of itself, it is interesting but it is not clear how it fits in, and I will suggest in a few moments how it can fit in. It comes right back to that general theme - the value added - the way you can bring yourself ahead and put yourself in a very competitive position is by integrating the sources, the detectors and the various supporting systems that would need to perform that integration.

A good example of a breakthrough that has just emerged at Rice University is a xenon fluoride excimer utilizing the C to A transition. Normally the B to X transition is the one which is tapped, and that's in the ultraviolet, I believe it is 3500 Angstroms, and through appropriate seeding of the mixture, with argon, krypton and some other black art chemicals, Rice University has been able to lase in the blue-green. You are still at about 1% efficiency which is standard for a xenon fluoride laser, but the key here is that it is a tunable excimer laser in the blue-green, from about 4500 Angstroms to 5200 Angstroms -- a spectral bandwidth of something like 600 Angstroms. Very important. Why? We within the Department of Defense, have been trying to come up with a blue-green laser for submarine communications for quite some time. And a tunable blue-green laser is ideally suitable, it is very suitable, for lidar systems provided that you have the right detector systems.

So, yes there is a great breakthrough in a tunable blue-green excimer laser, but unless that is coupled with a detector system, it will just sit there and be a wonderful innovation that people will applaud, but it just won't go anywhere. So you have to couple it in in the context of a system. So I keep on coming back to that very point because I think it is a very important point.

Cornell has come up with an interesting breakthrough. They have come up with a way to utilize insitu-Raman spectroscopy, when they grow their gallium phosphide and aluminum gallium phosphide thin films. And this makes the interfaces very very good, much better than before. And what it is boiling down to are visible lasers, solid state gallium phosphide visible lasers, at about 6800 Angstroms, specifically it is 6840 Angstroms, and 1:4 watts pulsed. Now, this is much better than achieved so far, and the previous record by the Japanese (and they are great at this, I can't speak more highly of the Japanese), was 200 milliwatts and this now has broken that barrier. And the threshold current density for this particular type of laser is about 1000 amps per square centimeter rather than 1700 amps per square centimeter. The efficiency is 56% rather than 37%. A great gadget, but how does it fit into systems?

Well, AT&T Bell Labs has been looking for a laser like this for their communications systems, their networking systems. This type of laser may be very suitable, may be the one that this (I mentioned this small company, Quantex and the optical storage system), needs to write onto its optical memory. You never can tell, but the match-up may be perfect. You have to bring these people together.

Another example of a interesting breakthrough in laser research, solid state laser, is the first room temperature CW gallium arsenide laser on a silicon substrate, and this was achieved at the University of Illinois. Their innovation, if you will, is by utilizing dry rather than chemical etching techniques. Now why is that important? Well, I'm preaching the choir, that we are approaching the Holy Grail, gallium arsenide on silicon laser systems at room temperature is something that we've been looking for for quite some time.

DETECTOR TECHNOLOGY

So, there is a lot of research going on out there, and it is all open. It is loaded into databases, it's there, you can access it but you have to spend a lot of time going through it and separating out the wheat from the chafe. But though breakthroughs, and I come back to this point again, by and of themselves are good but the value added is when you couple them and integrate them with other breakthroughs in totally different areas -- detectors for example. There have been a number of breakthroughs in detectors which couple very nicely with some of these laser systems. Example - the work going on at Kodak and at the Jet Propulsion Laboratory on iridium silicide detectors, which are CCD's, and they typically work at 9 microns, the LWIR. Typically the efficiency of those things is about 1%. Well there are some ideas, and they seem to be panning out, to bring the efficiency up to the theoretical limit for these things -- that is up to 10%. It's not there yet, but it is getting closer.

Platinum silicide, which is the SWIR range, 2-5 microns. Those detectors are beginning to make very large arrays, and they couple in very nicely with some of these sources that are being developed. Superconductivity. Although not so much in the laser area, in the detector area - there are some very, very good breakthroughs, wonderful breakthroughs, not too much in the yittrium barium, copper oxide high-temperature superconductors, but the fact that these types of superconductors, the rare earth as well as the standard materials (niobium nitride) are sensitive to IR radiation, very sensitive. In fact, the niobium nitride because it has a bandgap of about 500 gigahertz is a perfect detector for submillimeter wave applications.

Submillimeter, I realize I am not talking about lasers when I am talking about a 500 gigahertz or hopefully someday a terahertz source, but at that wavelength range you are very close to being able to capitalize on the tight bandwidth, the ability to carry a lot of information, the compactness of laser systems. We are talking about little pixi antennas on a chip with submilimeter wave systems. But you can capitalize on the various radar techniques which are extremely useful for various imaging systems. Why? Because radar techniques are based upon detecting electric field. You can regain the phase --you keep the phase-- whereas when you are dealing with lasers, unless you are dealing with very good lasers, you lose the phase information. (I emphasize unless you are dealing with very, very coherent lasers).

So, the importance of these new superconductors is that we may be able to extend the radar-like technologies. In other words, you are detecting electric fields rather than energy, up into the infrared domain. And you can imagine utilizing radar techniques in the infrared domain. The point I am trying to make here, is that there are many, many areas of detectors, sources, algorithms, networking systems, which all tie together. Which is why a meeting, such as this one here and SPIE in general, is so very important. It brings all these different people together, which approach problems in many different ways.

TEAMING TECHNOLOGIES

What we look for, certainly in my office, is given a particular mission (and that mission is relatively well defined in the context of what I am talking about), we look for teaming arrangements. We look for an entity which is able to come with breadboard products which can be integrated into that product which fits that mission. Yes, we are interested in the components, but we will try and do if we see an interesting component, we will try and match it with someone else's capabilities. In many case, at least initially when we tried to do this, there was some, not resentment so much but there was a hesitation in various companies wanting to work together like that. Everybody thinks they have something better than the other one. And it continues to be very difficult to convince Company A that has Technology A that if they work with Company B that has Technology B, the whole is much greater than the sum of the parts. And, both sides profit. That is beginning to work out in some of our electro-optic imaging systems, information systems if you will, that couple the laser marketplace with the detector marketplace.

So, I urge all of you in this room, if you are not already familiar with the other end of the system, the detector systems, to become much more acquainted with those detector systems. We have a lot of companies out there, companies like you, that are dying to meet you and both sides will profit tremendously.

I think that is about all I have to say, and there may be a number of questions. I can give you specific tips on where to go, off-line, or you can call me, or you can ask these questions here. The database is not with me, I have some information in my head, but it is hard to keep track of all of these things. The message I did want to deliver today was integrate. Find groups and businesses which are working in different areas which can tie together your technologies. It sounds like preaching the choir, it sounds obvious, but a lot of companies don't do that. In fact, many companies don't do it at all, they just try and stick to their own niche.

AUDIENCE QUESTIONS

Q: IN A PREVIOUS INCARNATION I INVESTIGATED WORKING WITH THE SBIR ON A PHASE I OPERATION. THE $50K IS ALMOST NOT WORTH IT. BY THE TIME YOU DEVOTE YOUR PEOPLE'S TIME, WHATEVER EXTRA RESOURCES YOU HAVE, AND THEN WAIT HOPING FOR THAT $50K, AND THEN DO $50K WORTH OF WORK WHICH IS ALMOST NOTHING, AND THEN GET AROUND TO THE CARROT WHICH IS PHASE II, ... THERE ARE A LOT BETTER WAYS TO SPEND YOUR ENERGY.

A: Well, the $50K, ... it depends upon the agency you work for. We have just received two or three hundred proposals for the Phase I. Those proposals are very, very short, and typically you can write one of those proposals in a couple of days. They don't have to be long, involved proposals. The key is that you have something which can be commercialized, and how do you know that it can be commercialized? Well, you have to deal directly with the program manager. The program managers will be evaluating those Phase I proposals. Their programs typically get cut - not the SBIR, the SBIR does not get cut. These program managers, they may have a program with line money. And they lose that line money, so they are looking for a way to keep their program alive, and the SBIR money is a great way to do it because it comes every year, it never gets cut. And what you can offer to that program manager, and he or she is right in here, .. if you find that their programs are being cut (even if they are not being cut), is an ability to enhance his or her program area. Because, if you went to them right now and described your idea, they would come to us and basically pitch us for the $50K.

      You would get it very easily, if they are on your side, and it would require only a couple of pages of proposal writing. In fact, you can use the two months that you have with that $50K to write the Phase II, which is, again, not a very involved proposal. It is definitely not as involved as the type of proposal that you would have to create for a venture capital group, or Kodak, or Dupont. And the key is to know the people. In this office there is only one layer of management between headquarters and you. So contact them. Whenever you send a proposal into the government abyss, you are right - you never hear from it again, it just kind of disappears. But, get to know these people. Just pick up the phone and call them, and if they do not talk to you, then you can call me, because I guarantee I will talk to you. I promise you I will talk to you.
      Has anybody else had any problems?

Q: I'M FAMILIAR WITH TWO COMPANIES INVOLVED WITH SBIR. THE PROBLEM IS LIKE CHASING YOUR TAIL ... YOU GET SO FAR ALONG IN A PARTICULAR TECHNOLOGY, THEY REALLY DON'T HAVE THE CAPITAL TO PRODUCTIZE IT. TYPICALLY THE SBIR FUNDING MIGHT BE 5-10% OF THE FRONT-END COST TO BRING THE PRODUCT TO MARKET. IN ORDER TO MAINTAIN A CASHFLOW THEY KEEP GOING FOR SBIR, BUT PRETTY SOON THEY ARE LEVERAGED OUT, AND THEY CAN NO LONGER FUNCTION AND THEY ARE TRAPPED.

A: Okay, the program manager's resources for your Phase II, which is the $500K, which will never be enough for capitalizing, are the same program managers that have $10-$15 million lines with me. That is the money that they could use if they wanted to to capitalize your effort. If you have an SBIR effort with them, and it looks very promising, and it probably will, those are the same people that would say, 'hey let's move you into the mainline program. Let's transition you away from this SBIR, this seed capital, into the mainline program, and begin brass boarding and breadboarding. We will give you a contract for X-million dollars to begin brass boarding.' That is coupled with your Phase III SBIR. The Phase III SBIR is a teaming arrangement with a major hardware house, a major industry. So what is different about this office, is that we don't have SBIR program managers, and if you will separate major project managers. They are the same people. The person that you talk to in here, ... yes, he is handing little $50K contracts, but he is also handing out 50 million dollar contracts. And in some cases, One Hundred Million dollar contracts, not out of my office but through other components of the DOD. So, in that way, this is a little bit different. We try to do things differently here, and I can appreciate your problem. Its a dead end unless you have that wedge in the fiscal door. You have to have access to a big bag of dollars somehow and that is through the program manager. And we have tried to do that here.

Q: WE, AS A SMALL BUSINESS, HAVE RUN INTO SEVERAL PROBLEMS WITH THE SBIR'S. ONE IS THAT YOU PUT YOUR BEST MAN WRITING THE PROPOSAL TO START WITH. THEN HE SITS AND WAITS FOR ABOUT NINE MONTHS TO PHASE I FUNDING. THEN HE GETS THROUGH PHASE I, AND WAITS AGAIN FOR FUNDING. IT IS VERY TOUGH FOR A SMALL COMPANY TO WEATHER THESE DROUGHTS.

A: Yes, you can submit the Phase II the day after you have a Phase I award. And, nine months is a long time, but I am sure most of you have had some business relations with other funding entities, and nine months isn't bad. That can drag on for many more months than that.
    Nine months is long for us. We try and get it done within six months, recognizing that you don't have this backbone to lean on.

Q: WHAT I AM SUGGESTING IS IF THE PROCESS COULD BE SPEEDED UP, IT WOULD BE A BIG HELP.

A: Well, it is part of the checks and balances. I don't think we can move any faster than Kodak could in investing in your idea.

Q: WHERE DO THE RIGHTS OF THE TECHNOLOGY RESIDE?

A: With you. The Government typically requests a royalty-free license, in other words, please don't charge us a royalty for the inventions that you come up with on our money. But in terms of commercialization, that is yours.

Q: I HAVE FRIENDS WHO HAVE THESE AND THERE ARE PROBLEMS SUCH AS, YOU
     ARE AWARDED IT AND ARE TOLD YOUR SIX MONTHS STARTED THREE MONTHS
     AGO FOR PHASE I. THE GUYS WHO ARE GETTING IT, IN MY VIEW, HAVE
     VERY GOOD PROGRAMS AS FAR AS GIVING IDEAS A CHANCE. NOW WHEN YOU
     GET INTO PHASE II AND PHASE III, I THINK THAT IS THE REAL PROOF OF
     THIS PROGRAM, AND WE HOPE THIS KIND OF INTERACTION WILL CONTINUE.
     THAT IS WHERE, FOR PEOPLE IN THIS AUDIENCE, YOU ARE GOING TO SEE
     THE IMPACT -- REALLY THE PHASE III LEVEL. THAT IS WHERE YOU ARE
     GOING TO CUT THAT 10 MILLION DEVELOPMENT COST DOWN TO SOMETHING
     THAT FITS INTO YOUR DEVELOPMENT BUDGET.

A: And you will have a good idea how you stand with the Phase II even
     before you get it, when you are talking to the program manager. You
     really have to interact directly with the program managers, and we are
     trying to make that very easy for you. Now they may put you off
     because they are so busy, but it shouldn't be for long. There are only
     so many people, but if you ever do have any problems, I will answer the
     phone. I have to answer the phone, otherwise I wouldn't dare stand up
     here! I will respond, and they will respond. You may not get an
     award, but you will get a hearing.

INDUSTRIAL LASERS'
IMPACT ON MANUFACTURING TECHNOLOGY

David Belforte
President, Belforte Associates
Editor, Industrial Laser Review

The international industrial laser market. Dr. Levitt referred to this in his talk, and I think David Kales made some comments about it. It is in his annual report that there were some changes in the 1987 numbers. In Fig. 1, we see a total of about 2335 units. CO_2 - 1400; Solid State - 900; Excimer - 35; a total of 2335. Now the column for the 1988 market. Those are Belforte Associates look at what we see the market to be. Notice that these numbers disagree somewhat with what David Kales reports in his January issue. Keep in mind that we have looked at this a lot sooner (or a lot later) than David did. This thing went to press in November, and we were still looking at this in late November. We have reflected some changes that came to us based on the market in Japan, which had a little uptick at the end of the year. A surprise to all of us. It was an uptick in Japan and a bit of an uptick in Europe, and those of you who read the press noticed that Spectra-Physics reported that they had a rather good year, and Coherent General for example told us that they had a pretty good last quarter of their year. All of this information came somewhat after the fact when the Laser Report information was put together. So I am reasonably confident that within a couple of a percent or something like that, these aren't all that bad of numbers for projections for 1988.

Installations Worldwide vs the United States. I have changed this percentage to reflect a minor difference in the 1987 reporting, and there was a correction which had to be made because of a change in the reporting we received from Japan. We had looked a little bit more optimistically at Japan when we put these numbers together. We have backed away from our more optimistic view of Japan for the moment, even though there seems to be a bit of recovery in Japan. Maybe Joe Nagasawa will be discussing that for us later this afternoon. But we have moved our number back down. Notice the 2225 has the excimer laser sales taken out of it. That is the only difference. It should be 2325 if the excimers were put in. Which still says that the U.S. is looking at a roughly 38% share of this business.

With these numbers in mind, it is therefore my conclusion that contrary to our seeing a decrease in the industrial laser sales for 1988, basically it is a flat year. We are not going to see any growth to speak of. I don't really think we are going to go negative at all. I am sorry, I do disagree with David on that, and I think he and I have talked about this enough the last month that he feels a little bit more confident as I do, in the fact that we are probably going to have another flat year. Is that bad? Looking at the capital equipment business, it is not all that bad in my estimation. I think, when I talk about some trends a little bit later in this talk, you will see why I am a lot more bullish about the business perhaps, than other people might be.

Figure 1

The Industrial Industrial Laser Market

(Units)

Type	1986[1]	1987[2]	1988[3]
CO_2	1300	1400	1400
Solid State	890	900	825
Excimer	36	35	75
Total	2226	2335	2300

Source: [1] Laser Report – January 1987

[2] Laser Report – January 1988

[3] Belforte Associates

I made one minor change here just to prove to you that we are trying to remain correct, and that is I have decreased the 1987 units % down to 37% and the European up by 21%. I am scanning through the information we are going to hear about the European market, and I noticed that they have a little higher percentage there. I am willing to agree on that. The one bright light in overall international sales in materials processing in 1987 was Europe. They really did have very, very good growth over there and it kept the entire business propped up in the face of a severe recession in the capital equipment sales in Japan, and kind of a blah year here in the U.S..

Now, we referred earlier in Industrial Laser Review to some work that is done by the Prognos research organization from Switzerland who has been looking at the markets. You have probably seen this information in both ILR and Laser Focus, but I put it up (Fig. 2) to show you that totally independent people are looking at the same kinds of information that we look, have come to about the same conclusions we have. That, no, CO_2 lasers have not matured to the point where they are going to see decreasing sales in the industrial marketplace and, indeed, we all believe quite strongly that there is growth. And I must

Figure 2

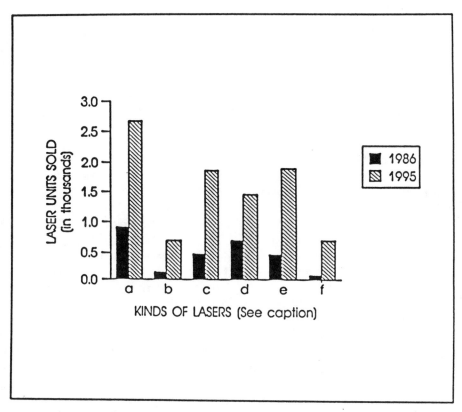

The graph shows a changing world market for laser systems used for industrial material processing. The bars represent the sum of the markets of the U.S., Europe, and Japan, in thousands of units.

Key: a—CO_2 lasers, 200 W to 1.5 kW; b—CO_2 lasers, 1.5 kW and more; c—solid-state lasers for marking; d—solid-state lasers for microprocessing applications; e—solid-state lasers for welding, cutting, drilling, soldering; f—excimer lasers.   Source: Prognos

say, these Prognos people did a pretty good job on their survey. I know people that were interviewed by them, and I myself was interviewed. I know they have done a pretty good job of digging it out, and it must be a quality report because they are asking several thousands of dollars for it. I don't know how many people have bought it, but we had a look at it and we are kind of impressed by their methodology of getting the information. We may differ with their numbers out there, that far out in advance - 1995, but at least they, along with us, look for a continuing upward growth. And they also see, basically, about the same things we see in terms of who is going to get what share of the business. And they see that out past the 1990's that Japan is going to become the dominant consumer of industrial laser systems. These are systems, remember, not devices they are talking about here.

Regardless of whether CO_2 lasers this year go negative and therefore industrial laser sales are negative or flat or are going to enjoy low double digit growth, there are people in the industry that say even if we have a small growth, the size of the market is not big enough to support the players in the marketplace. On Friday I had a long interview with the president of a major supplier of industrial lasers and systems, who said to me, for God sakes next Wednesday, only give out bad news. You guys have been giving out good news for too long. That means more players in the marketplace. Will you either fib about it and make it look as black as you can or tell the truth. So I said I was going to tell the truth, and the truth is that we think it is kind of a flat market. I am sorry we can't make it look negative for you so you won't get any competition.

A lot of people have said, and you have read about it in the publications, that there is going to be a shakeout. David Kales reported in Laser Report last year, a big shakeout is coming. Possibly so. But I would like to share with you some thoughts I have on it. These are the people who responded to the 1987 questionnaire for Industrial Laser Annual Handbook. These people have told us, and we qualify them by looking at their specification sheets, that this is where they perform in the area of industrial laser goods and services. Now there are other companies out there, but these are the ones that responded to our survey. So we know where 600+ of these people are and what they do. There are approximately 200 people who say they supply lasers, lasers to OEM's, or they are the integrators themselves of laser systems. There is no double counting up here. So, if we are going to have a shakeout, if somebody has to get out of this business to make this market more profitable for the people who are in it, who is it going to be? Let's take a look at it. Of these 200 people who play in this portion of the marketplace, 30 of these companies are large, dominant suppliers of lasers and laser systems, and because of their multi-divisional capability, the fact that they are divisions of large international corporations or large corporations, it is highly unlikely that they are going to abandon their marketshare. They could withstand a recession in the machine tool industry more than other people could. So, if we take these 30 people and say, they are going to be here in 1988, it turns out they represent about 75% of the business. In Industrial Laser Annual Handbook we are going to make a list of them, but you know who they are: Coherent General, Spectra-Physics, Mitsubishi, Amada, Ferranti, these are the companies that are going to be around because they are in other things and they can carry this activity for a while.

Then there is a second tier of people. About 30 of these companies I call niche marketing people. These are the Teradynes of this world, Chicago Lasers, companies like that who focus in on a certain part of the market and are pretty good at that part of the market. Although they may suffer swings in what goes on, they will still probably be here. This category has about 15% of the business. So out of 200 people, we have 60 people who have about 85% of the business, maybe as high as 90%. So 140 companies are sharing 10% of the business. I admit that amongst these 140 companies, there are some very interesting companies with good technology and some great ideas. But frankly, being very brutal about it, if they were to depart the scene tomorrow, it is not going to have a significant impact on the survivors. I don't believe that it is the little companies, the guys that are just getting going, that make life difficult for the big guy. I think it is other big guys making difficult problems for the big guys.

So, if we have a shakeout, and if that shakeout comes via some mergers - for example we have already had Siemens buying Rofin-Sinar; we have Ferranti buying Sciacky we have Mazak teaming up with Nissho Iwai. There are a whole bunch of these things going on, so maybe some of these numbers do drop, especially in the systems integrator thing a little bit. But I think the bulk of these 60 guys who do 80-85% of the business will be here, and if the smaller guys do merge or disappear, I don't think it is going to have a significant impact on market share.

I just wanted to take a look very briefly at where these lasers are being used (Fig. 3). We try annually to talk to people, the suppliers who will share with us information about where are your lasers going? Those of you who were here for the last seminar, you will find the percentages haven't changed much. They really haven't because cutting and drilling remains a rather big segment of the marketplace. Welding and soldering a little bit less, and then we move down to marking and scribing and trimming, etc., and it varies from country to country. Japan has always been strong in cutting. Europe is quite strong in cutting now, whereas the U.S. certainly can't match those kind of results, but they tend to be higher in other areas. So, we haven't seen any significant shift at the moment through 1987 in where the industrial lasers are being used. What we do see is a shift happening in 1988 in the welding section, which I will talk about in a minute.

We received responses from 300 companies from our questionnaire that we send out. We send them out to about 275 or 300 of our old company names in our directory, and 300 or 400 new ones that we get each year, and we accumulate about 300 - of which about 150 take the time to share with us their thoughts on trends. Now admittedly, we get this information in the summer. We had most of this last August, so it is my function as Editor of Industrial Laser Annual Handbook to put them together for thoughts in the introduction for this years' book, the 1988 book. We just got through doing that, so I would like to share with you what your peers have told us look like trends for 1988 and a little bit beyond. Remember they told us this in August, so very little of it is financially oriented.

Across the board uniformly, selling prices will decline based on manufacturing cost declines something on the order of 3% per year for the next five years. Excimer lasers - sales will increase at a higher rate because of increased reliability in the product and reductions in unit cost. Full production use of excimer lasers will begin in 1988. Pilot plant production that is going on now will begin to be full production operation. We see the beginnings of very significant growth in the excimer laser market for the next five years. We think 1988 will be the transition year when some industrial units, true industrial units, will go into production, perhaps later this year. But well into

Figure 3

% Distribution of Industrial Laser Processes

(Installations through 1987)

Process	U.S.	Japan	Europe
Cut/Drill	36	60	47
Weld/Solder	24	20	20
Surface Treatment	1	1	1
Scribe/Trim	21	12	20
Mark	15	5	7
Other	3	2	5
	100	100	100

Source: Belforte Associates

1989 we will begin to see these pilot plant operations taking place. Where? We think micro machining of polymers metals and ceramics are exciting prospects for excimer lasers especially when you look at the European funding for excimer laser development in the high/average power, high/peak power excimers is going on over there.

You are going to see later on today in a presentation, some of the BRITE and Eureka money that is being spent on excimer laser development in Europe. It is millions of dollars being spent over the next few years to create industrial product and the applications for those. Most of these programs are four-year programs.

The supplier industry that produces the products that we just talked about - all these guys that are out there, these 200 people, are going to become more customer oriented. Now you are looking at me strangely, but your customers do not think you are customer oriented. If they are OEM customers, they know you are not customer oriented. If they are end-user customers, they agree that you are not customer oriented. They really want to see some better selling going on. They want better sales people calling on them, they want more people listening to their problems and trying to put those problems into perspective. Along the same line they are telling us that the systems that these people are going to be building for them are going to be built with a machine tool mentality. They see that happening now.

Another thing that will happen in 1988 is, we are going to see a surge in multi-kilowatt CO_2 lasers for industrial applications. Probably at the end of 1988 we will be predicting the highest percentage growth rate in this multi-kilowatt laser level. It may be delayed by about six months, it may be into 1989, but certainly in 1988 we will begin to see very large increases in sales of those products.

We are going to see more lasers in flexible manufacturing system (FMS) operation. FMS is becoming fairly common now. It used to be we talked about why hasn't it been implemented, we see it being implemented, and there is more and more laser based FMS coming on-line.

Marking systems. We received rather neutral and in some cases negative comments about the opportunities in marking in 1988 and beyond. We don't believe that is true. We think that marking has vast potential still available to it. Yes, there are perhaps too many people in the marketplace offering systems, but indeed that isn't going to stall the business. The business will grow. There are too many things that lasers will be able to do based on lower cost systems and more user friendly marking systems.

In general then, when we look at those trends, which I think are all very positive trends, we already see the beginnings of some of them happening because remember, we were told about these in August, and here it is January already. You can see why Industrial Laser Review and myself, and other people who look at this a little bit more closely, are somewhat more bullish about the market. We do believe, however, that we are in for a period of single digit growth, low single digit growth, in the industrial laser community for some years to come. There may not be that many people participating in the industry, and therefore the percentage share and profitability for those companies who do remain in the industry will be better. But I think you will see that when we are dealing with 3-, 4-, to 5-thousand annual unit sales out around 1990, the single digit growth isn't all that bad because it is very consistent with the machine tool industry growth.

So trying to put a better view on the rather negative news that Dr. Levitt showed you this morning, I think that we are looking at a little bit more positive news in terms of industrial laser sales and I think that CO_2 laser sales are going to hold up because they still remain the backbone.

AUDIENCE QUESTIONS
Q: WHAT IS ABOUT THE REPLACEMENT OPTICS BUSINESS, ESPECIALLY EXTERNAL OPTICS.

A: Yes, we looked at this. Dr. Levitt and I looked at this last year because we wanted to make sure that our Industrial Laser Annual Handbook was presenting valuable information to our end-users, and we thought that replacement optics was a business that perhaps we ought to take a look at to see if it was significant. So we interviewed a lot of end-users to try and get some feeling about the business. And frankly, we were a little bit surprised that the employee, the operating personnel effect on external optics is rather minimal. Guys are getting pretty good with how they treat optics out there in the field, they seem to be getting much longer life, and based on that it is my impression that the optics business is going to enjoy fairly small growth in the industrial replacement business because they are getting much longer life. As we get perhaps into single digit growth, and we are talking about hundreds of units more out in the field, then that market should continue. But I think in talking to people at II-VI and Laser Power Optics and others earlier this fall, I got the impression from them also that people are learning how to handle optics out there. In the higher power end of things, the metal optics are really taking off and have begun to replace transmitting optics, and they are just getting phenomenal life. People are learning more about these things. They are less sensitive to optics, and I think the suppliers do a better job of putting them in.

Q: YOU COMMENTED THAT EXCIMER USE WAS GOING TO INCREASE BECAUSE OF TWO REASONS -- 1) RELIABILITY AND 2) LOWER COST. I CERTAINLY AGREE WITH THE FIRST PART, BUT THE SECOND PART MAYBE NEEDS TO BE CLARIFIED.

A: Yes, thank you. Lower cost for power delivered.

Q: THE CAPITAL COSTS FOR INDUSTRIAL LASERS IS 25% TO 50% ABOVE COMPARABLE COSTS FOR SCIENTIFIC LASERS. THE OPERATING COSTS MAY BEGIN TO COME DOWN, BUT NOT IMMEDIATELY.

A: Yes, but also the unit costs for the product throughput, for power delivered, actually will decrease. And that is what it is based on. We are too caught up with the idea of how much does a unit cost, and the end-user talk about how much per piece part when I turned it out, and that is what I really meant. You are right on the operating costs.

Q: ASIDE FROM MEDICAL AND SCIENTIFIC MARKETS, WHICH INDUSTRIES DO YOU SEE IMPACTED BY EXCIMER LASERS?

A: Well, you know you want to say micro electronics, but frankly that is what it is. I am like you or most of you people in this room. I guess you are just fascinated by what you are reading about micro machining and the opportunities to make these little tiny motors, etc. Help me out a bit, I am not exactly sure where they fit. I put them in the area of micro electronics, semi-conductor processing. There are people here who are much more expert than I am on what the specific industry is, but I will tell you this, it is an industry that we have not been reporting on now because we think of industrial lasers sales as being the cutting people, the welding people, automotive, aerospace, and we have been negligent in bringing those people under our umbrella of industrial lasers. We intend to correct that in 1988, and start taking a closer look at that so we can report on this much more closely, about this part of the industry. I guess that is where it would sit. Perhaps people from Quantronix or other places could tell you better about what these industry segments are called.

Levitt: I would like to add one comment to the question about the optics market. We did a survey last year based on about 10,000 consumers of industrial laser products that we can identify, and within a group that indicated that about 80% of them would be purchasing industrial lasers. I think was close to something like 20%, said they might be purchasing optics. so those are very rough numbers, but it gives you some scale of comparison.

MEDICAL LASER MARKETPLACE
A YEAR OF ACQUISITION AND REORGANIZATION

Michael Moretti
Editor, Medical Laser Industry Report

Depending on how you are positioned in the industry, it could be a good year for you, it could be a good next few years. There are some very nice projections into 1990 for certain areas. In other areas such as the ophthalmic market, certain segments of the ophthalmic market, there is notable decline, so it might be a time to reposition.

First I am going to present an overview of the medical laser market (Fig. 1). I am going to focus in on specific segments in growth areas. All of my data comes from original research I have done for Medical Laser Industry Report, my monthly newsletter.

Figure 1

Worldwide Medical Laser Sales

(Units Sold)

	CO/2	YAG	Ion
1987	1400	1200	1900
1988	1600	1400	2150

Source: Laser Focus/Electro-Optics magazine.

Translates into 1987 approximate dollar values of:

CO/2	=	$60M
YAG	=	$80M
ION	=	$65M

(Ion laser numbers do not include sales for diagnostic applications.)

Laser Focus Magazine has taken a slightly different approach to quantifying the market, and in their latest Jan. 88 report, they broke it down into units and I extrapolated, or made some guesstimates from those units, of what the sales were. So here is what I would say are some conservative guesses, the dollar amounts at the bottom of the table.

The first segment I would like to target would be the ophthalmic ion laser market (Fig. 2). I have broken it down into units in 1987, units in 1990, and what the values are. In the ion area, it doesn't look good. The average selling price now, you can see $35,000. There might be replacement markets, if users decide that it is too costly to replace the tubes, it could be that they will upgrade to new units, possibly even combinations of argon and YAG lasers, or even tunable dye lasers, with a variety of wavelengths, and therefore, more applications.

Figure 2

Ophthalmic Ion (argon/krypton) Laser Market (U.S.)

	1987	1990
(units)	450	350
(value)	$17M	$14M

- average selling price = $35,000

- replacement market may materialize as tubes fail; large installed

- tunable dye laser will be choice for upgrade in some cases

The next segment is the ophthalmic YAG laser market. Again, units 1987, 1990 and the values (Fig. 3). You can see the decline. In the ophthalmic YAG unit market unit sales were about 450 in the previous year (1987), my guess is 400 by 1990 -- a slow shrink. The 1987 dollar value, (and I'm being conservative), was 15 million declining to 13 million in 1990. That is based on the average selling prices of these lasers. The average price is about 30 thousand dollars in the YAG category.

Figure 3

Ophthalmic YAG Laser Market (U.S.)

	<u>1987</u>	<u>1990</u>
(units)	450	400
(value)	$15M	$13M

- office and surgicenter sales far exceed institutional

- average price is $30,000 -- stabilized

- combination YAG/argon laser is well received

One new laser which entered the market this year was the combination YAG/Argon, which got a nice reception at the Academy (of Opthalmology) meeting in Dallas, and could mean some new life for the industry, although I don't think a lot.

Just by way of example, you can see here (Fig. 4) a list of the companies that were active in 1984 vs the players that are left in 1988. The 1988 list are mostly strong, but I would expect two or three more to withdraw within the next year.

Let's talk about price trends. The average price of an ophthalmic laser was 55 thousand in 1984, prices declined to 40 thousand by 1987. Towards the end of this year some of the industry leaders, survivors (namely Cooper and Coherent), raised their prices slightly in certain categories. Overall I would say prices have stabilized and the price war, if not over, at least they have called a truce.

I think it is more meaningful to talk about trends than dollars actually. Again, manufacturers are leaving. The argon and YAG are flat and declining. Institutions are saturated. Private practice sales were promising but never materialized. Dye laser growth, there might be some growth as replacement for argon/krypton lasers. Two very hopeful areas -- excimers for corneal sculpting and a variety of lasers possibly for a tongue twisting technique called photophacoemulsification, which uses a laser in a novel way to remove cataracts.

For the corneal sculpting market - I have made a rough attempt to quantify what the market is and what it will be. It is estimated, 2000 sites for these units. Final systems are going to be expensive, 3 to 4 hundred thousand dollars. Depending on the efficacy, we just don't know what percentage of the population will ultimately be treated. It could be from anyone who wears eyeglasses to people with serious corneal defects, or possibly blindness.

Figure 4

Ophthalmic Laser Companies

1984	1988
AMO	Biophysic
Biophysic	Coherent
Britt	CooperVision
Cilas	Gish
Cilco	HGM
Coburn	Lasag
Coherent	Meditec
CooperVision	Nidek
Gish	Sharplan
HGM	Zeiss
Humphrey	
LFM	
Lasag	
Lasermed	
Lasertek	
Marco	
Medical Lasers, Inc.	
Meditec	
Nidek	
Sharplan	
Storz	
Zeiss	

I have some highlights from the American Academy of Ophthalmology meeting in Dallas, that was a couple of months ago, and it is sort of is the pulsebeat of the ophthalmic laser market. There is a mystery video from a company called Healthtek. They emerged from nowhere, and they had what I would call a nothing kind of a video. But they made a lot of promises and the word photophacoemulsification was dropped during the course of the endless loop video. Upon questioning, we find that they will have a laser-based photophacoemulsification system late this year or next, and there are no details on what sort of a laser is used in that system. It is an exciting idea though.

Another interesting item - Coherent's portable YAG. They got a lot of press on this for treating patients in Third World, unfortunately they don't have much money in the Third World. But it does fit in a suitcase, and you could easily take it on a plane. Speaking with the Marketing Director of Coherent's Ophthalmic Division, he tells me that some local domestic ophthalmologists have decided that it is right for them. They can travel on their vacations and still take their YAG with them, that sort of thing.

So here is a similar model from Gish, and basically the same thing, although many people indicate that it is a bit shoddier. It also fits in a suitcase and performs the same function.

Excimer Laser from Summit Technology. This is a sort of laser that is being used in corneal sculpting research and also, according to the company, although I haven't seen any evidence, in photophacoemulsification experiments. Photophacoemulsification is using a laser fiberoptically delivered to ablate the cataract lens rather than the posterior capsulotomies that have been famous. This goes in and directly gets the lens. Competing technologies are surgical removal and ultrasonics. Cooper Vision also has an excimer laser for research although this one is larger and more expensive than the Summit model. It is being used at a few places in the country for research right now, and it is a giant monster of a system.

Taunton, a new company has emerged. They are starting out like a class act, they have patented the concept of corneal sculpting. Leading ophthalmologist Francis L'Esperance is chairman of the board or some notable position, and they are pretty much trying to monopolize the corneal sculpting business even before it gets off the ground. Aside from any clinical efficacy issues, there are already legal issues, patent issues, at stake in a market that people are very excited about.

In the area of laser-based diagnostics for ophthalmic applications - this is the long awaited Rodenstock scanning ophthalmoscope. It performs a variety of diagnostic functions on the human eye, and while I don't have a price on it, I know it is being used perhaps in one or two key clinics already to screen patients for eye diseases.

Another notable laser that was developed actually for the ophthalmic market a couple of years ago, represented here by Meditec, a tunable dye laser. It has taken off to some degree in that profession. The advantages are instead of having one wavelength let's say with argon or two with argon/krypton, they now have a variety of tunable wavelengths to treat various eye tissues with the tunable dye laser, so it is sort of the fanciest thing around. The absorption peaks of the dyes correspond to key ocular tissues. You can predict exactly what happens to an argon, the 488 or 514 wavelengths on tissue, the 577 wavelength on tissue, and the 630 on tissue. It is the hope of some of these dye laser manufacturers that not only can the laser be used in ophthalmology but using fiberoptics, it can be piped around the facility with local control panels and be used in such things in later Photo Dynamic Therapy (PDT) research. We will talk a little bit more about PDT research later.

Here is another diagnostic instrument developed by Computed Anatomy. It is a start-up of, I think, three ophthalmologists. It is a very sophisticated computer driven system. It models the cornea which gives it a lot of applications. The company really hasn't explored all of the uses. One suggested use would be to drive a real time monitor of the corneal sculpting process. And, in fact, there may even be negotiations between some of the R&D companies that are looking at corneal sculpting and Computed Anatomy. It would be a nice marriage I think, because typically the laser manufacturers are weak on the software side, and this could be where they come together.

The corneal modeling system is just a straight video image of an eye, and it overlays these rings. It uses that information to image process and, I believe, if we were clinical people we would know that this is an astigmatism.

You can image and model three different representations of that condition and do all sorts of unique things with this system although it is still under development.

THE SURGICAL MARKET

Let's move on to the surgical market. My guesstimate is that we are currently at about 125 million dollar sales worldwide. This is lower than a lot of people think, but I am more realistic after studying this day and night for the past couple of years. Growth rates vary widely. Every hospital seems to want a laser if they don't have one; there is a good potential in the office especially for gynecologists. There is a new breed cropping up of laser rental people. It is like video rentals, and they could put the movie houses out of business ... but that remains to be seen. YAG popularity is growing, especially urology. Prices have stabilized generally, as I said earlier.

More specifically, I narrowed it down to the CO_2 laser market. In 1987, a 70 million dollar guesstimate. In 1990 I see that doubling. This is the hottest market, and the most useful area will be in that range - 25 to 60 watts. These are the current leaders, Sharplan Coherent/Xanar (which they acquired). And you could argue in the lower power models that Xanar actually is the market leader. I'll leave that up to those two companies.

This is the line of Sharplan CO_2 lasers, they have some very impressive technology, well thought of by users. Model 1020 is a 20 watt portable. This is very representative of one of the fastest growing market segments for CO_2's. This is the kind of laser that would be used in a surgicenter or in an office, rented, wheeled around, move it, throw it in your trunk. Very useful.

Surgical YAG laser market is not so big; for 1987 - 30 million dollars, for 1990 - 40 million dollars. Very small. The greatest potential seems to be in urology, but there aren't that many urologists. In contact laser surgery where it is targeted at general surgeons mostly, and there are a lot of general surgeons, the problem is penetration. They haven't been able to convince general surgeons to drop their scalpel and pick up a contact laser scalpel. (I'll show you some of those later). If they ever do, it could be a significant market and change all of these figures. At any rate, there is no longer a big demand for 100 watt YAGs. If you are working in the contact mode, or if you are a urologist, certainly a urologist doesn't need much more than 40 watts, and a general surgeon doesn't need much more than 75 watts.

Also promising applications in novel areas such as kidney stone removal - we all know about Candela's success in that area with their dye laser -- well if you had a much less expensive CW YAG laser or pulsed YAG laser to bust up those stones with a fiber delivering the beam, I think you would be in a very competitive situation. In fact a couple of European companies are developing those technologies right now. A variety of contact tips are available from Sharplan, and you have varying geometries depending on the application. Similarly,

Surgical Laser Technologies uses tips that Stephen Joffe and Cuzono in Japan invented. Joffe brought them to America, and tried to popularize them. In fact I think there is even a patent dispute now over the use of these tips by other companies. Energy distribution depends on tip geometry. The various distributions mean the surgeon can select one based on exactly what he wants to accomplish. You might want to go shallow-deep, narrow-deep, that sort of thing.

In terms of manufacturing, there are some key points. A poor optical design leads to poor efficacy in operation. The perfect optical design gives you perfect distribution of that energy.

Here are some of the key attributes of contact laser surgery that many people think will make it popular. Depth of tissue necrosis, that's the main thing. When using a bare fiber YAG you don't know how deep you're coagulating tissue and you don't know how much tissue you are destroying beyond the area you want. You can see there is a significant different between contact and non-contact modes.

Smoke production. A lot of nurses and doctors are getting worried about the hazards of smoke produced during laser surgery, especially when they are operating on condalomas and God knows what else at this point. They don't want to breathe too much of the smoke. So there is another attribute in favor of contact surgery, less smoke.

Blood loss. Not only does this show blood loss, it gives you an idea of the amount of power, the difference in power, that you need to operation. So you no longer need a 100 watt YAG, you can have a 25 watt YAG accomplish the same purpose, using contact tips because it delivers the energy to a much smaller area and requires less input.

THE FUTURE

For sales growth, I see 15% growth for YAG lasers in the near term. CO_2 lasers, I will break that down into hospital and office. Hospital is a much more mature area of laser sales, I see 10%. 25% for office sales, although some people have gone as high as 35%. But I will stick with 25%. The office market adds up to about 15% of all current sales.

Trends. There is a move towards the 50- or 60-watt CO_2 and YAG lasers. Fifty-five thousand dollars is about the average budget at this time when I talked to users. They don't want to pay more than that at this point. They are tired of buying the one hundred to one hundred fifty thousand lasers that never prove to be what they were talked about. Air cooled, plug it in the wall, move it around, sealed tube, they don't want to mess with the lasers. They want it to be like any other instrument in the hospital.

And, lastly, specialized delivery systems that allow a lot of different specialists to use the laser.

User needs. It is a little foggy here, but definitely some needs have been identified and meeting these needs could mean increased sales or success in various niches. Mobile, compact, reliable - electrical efficiency (they want wall plug units), they want air cooled, they want a combination laser, CO_2 and YAG for cutting and coagulating. Imagine the logistics problem of trying to use two lasers in the same OR. You can't do it, but you can do wonderful things if you have both in one unit. There is a Japanese company, at least one Japanese company, and Cooper Laser Sonics also has a combination laser. They are both quite expensive. The price has to come down - you saw what they are willing to pay now, 55 thousand dollars. The price has got to come down below one hundred thousand dollars before anybody is going to buy something like that. But the key thing here is multi-function lasers. In the economic environment, in hospitals and health care industries, they have to do a lot of talking to get a budget for a laser, and they can go much further if they can claim to the administration that it is multi-functional, it can be used by a lot of people.

Tunable lasers. There is some research being done by a few companies, and I believe this is a very, very hot potential area because if you could have a tunable solid-state laser, tunable to the degree that it could perform cutting and coagulating, and in power ranges from 25- to 70-watts, you would have a winner.

Other market opportunities are outside of lasers in the peripheral products. Delivery systems, surgical instruments, smoke filters, angioplasty - everybody knows about angioplasty - photodynamic therapy - everybody wonders about photodynamic therapy. Let's look at these more closely.

For example, anti-reflective laser instruments are a very popular accessory. There are several companies supplying them.

Smoke filters. This can be plugged into the existing OR in-line evacuation equipment, and it costs about $5.00 apiece. Here is a much more sophisticated model which is far more effective, and has been very popular, especially with CO_2 laser users. That's the unit made by Stackhouse.

Fiberoptics sensor markets segment. These are very real technologies under development, if not already implemented. Blood gas monitors are already used, temperatures monitors are already used, diagnostic spectroscopy is primarily a laboratory concept right now, but it has a lot of potential if you can identify the tissue you are working on in areas such as angioplasty and other areas, you would have a pretty good product. The worldwide market for medical users of optical fiber for 1986, totals 150 million dollars. Sensors will grows enormously, again, this is according to Kessler, for a total market of 732 million dollars by 1995.

Let's move on to laser angioplasty markets. By 1990 it is projected that the coronary angioplasty market, all therapies whatever they use, will be about 600 million dollars. Peripheral angioplasty market, that is 40 million dollars by 1990. Now the first year any laser angioplastys were really sold, and long with disposables, was 1987. Now only a few have been sold, but numbers have already reached 10 million dollars. And let me tell you, hospitals are clamoring to be the first on the block to have an angioplasty system. In the competitive marketing atmosphere among hospitals now, having something like laser angioplasty could give you the edge. So as a result, Trimedyne is running to the bank and all the other companies are running to catch up with them, companies like G.V. Medical, Vaser, and a handful of other companies. By 1990 I see sales of laser products for peripheral and coronary applications reaching 75 million dollars. I think that is reasonable. A lot depends on the regulatory status of these products, but I think that is reasonable given the timeframe, and what I know about the regulatory process.

By then, however there probably will be at least 12 strong, if not strong, active, companies. And from 4 to 6 approved products competing in the peripheral coronary markets. Still, you have to see those as somewhat different markets initially. A product will get approval in the peripheral application before it will get approval in the coronary application.

Here is an example of an existing system that is under review with the FDA right now, it is probably right behind Trimedyne in terms of getting approval for use. The system is from G. V. Medical in Minneapolis. Basically you have sort of a hand held catheter control unit which is a balloon assisted laser angioplasty system. It operates by the balloon expanding and blocking the blood, and the laser operates to remove the blockage. The balloon deflates and releases the blood. While the balloon is up it also acts to maintain pressure on the wall of the vessel and in that way it helps to open it even further, similar to straight balloon angioplasty that is being used now.

One last segment, if I have time, I would like to mention is what I would characterize as a sleeper. Photodynamic Therapy (PDT). For years, maybe ten years ago, people were making claims about the efficacy of PDT, and in the meantime several companies, notably Johnson & Johnson through their unit PhotoMedic have pursued FDA approval of a drug, the latest version being Photofren-II. Well, Johnson & Johnson lost so much money that the decided to sell it. It took them a year to unload. They finally have to a Canadian company by the name of Quadrologic Technologies (QLT) which acquired rights to the drug and all clinical data that had been collected. Now, QLT will reinitiate clinical trial aggressively to seek approval, and not only for that drug but they have advanced next generation drugs that bind antibodies and things like that, to target the drug. So they have a lot of ideas, but their first task is to get a drug approved. When they do that, I believe that as many others do, that PDT will be a significant laser market. It is just a question of how much money does QLT have. How soon can they get it done? The only promising thing is that, I believe, Johnson & Johnson was actually in the third stage, if I am not mistaken, of the clinical trials when they bailed out. So, it may not be that much further to go. I don't have any numbers on that market, although there are people selling market studies.

In summary, there are current opportunities in certain market niches, as you have seen, like YAG and CO_2 lasers. Other opportunities are emerging rapidly in ophthalmic like photophaco and corneal sculpting; cardiovascular - both peripheral and coronary. Future areas of growth: Fiberoptics sensors, fiber delivery systems and all of those accessory products. I didn't mention a very significant accessory area and that is operating room disposables. Now there is a whole line of disposables that are just used with lasers, and a company by the name of Baxter Travinal has decided to jump in there and take the share, so right now they are about the only company that has a full line of OR disposables for lasers.

All of these market areas are analyzed in detail, month by month in Medical Laser Industry Reports.

AUDIENCE QUESTIONS

Q: ON YOUR SLIDE ABOUT LASER ANGIOPLASTY, I MISSED THE DIFFERENCE BETWEEN THE 1990 FIGURE OF 600 MILLION AND THE 1990 FIGURES OF 75 MILLION.

A: The 1990 figure of 600 million quantifies the entire market. It has nothing to do with lasers, just all therapies. If you just take all the therapies that might be used. What I am saying is, in 1990 I expect 75 million worth of laser systems.

Q: WHAT WAVELENGTH(S) ARE BEING CONSIDERED FOR PHOTOPHACOEMULSIFICATION AND HAVE ANY BEEN APPROVED BY THE FDA?

A: The companies are very secretive, nothing has been approved. We can guess that they are diode pumped Q-switch YAG or possibly excimer; something in which they have extremely high energies, because you cannot have any heat generated.

Q: HOW MANY PLAYERS?

A: Three or four companies are involved.

Q: WHAT WILL THE IMPACT OF PATLEX BE?

A: I really don't think prices can be raised, the days of high prices are over. It has already affected Cooper. We'll have to see what happens to Coherent.

MILITARY MARKET
EXCLUSIVE CLUB OR MAJOR MARKET OPPORTUNITY?

Dan Morrow
Manager, National Defense Architectures
BDM Corporation

The topic the title is Military Market - An Exclusive Club or Major Market Opportunity? I think it really depends on your specific product. I know a lot of people view the military as being an exclusive club. I think some of the people who have had the opportunity to see Morely Safer and the 60 Minutes crew come into their office and talk to them about $600 toilet seat really wish they weren't members of that exclusive club. But I have worked in the military area, where I am kind of complementing what Dr. Ionson had to present earlier, except I am kind of on the other side of the fence. I have been a defense contractor for 11 years, so I have a little bit different point of view than the government does. But a lot of what I am saying tends to be a near-term market kind of approach rather than a longer-term sort of product development kind of information. But there certainly is, you are right, a lot of money in the government. This year's military budget is about $290 billion dollars, on the order of $80 billion of that goes into buying things. So there is a lot of money out there. What I am going to be talking about today is how much of that money goes into lasers. I am going to try to quantify it and also give a little bit of a qualitative estimate of where the trends are going, what is going to be the next lasers to be bought.

Anytime you do a study like this, there is a set of assumptions that you need to be aware of so it kind of scopes out for your what exactly the answer means when you get through looking at it. And the most obvious is that we have only looked at the military market. We are aware of the U.S. military market, we are aware of the rest of the world, but for today's presentation, we just looked at the U.S. Another important thing is to be sure when you talk about dollars that you talk about them in all the same year. If you gather a lot of data, you can't compare 1977 dollars with 1988 dollars, they don't compare. So the analysis has been rolled up so they are all compared properly. Another thing is that some of the information that we've seen published tends to talk about the marketplace in terms of the whole integrated end item. And we've taken care, with our data, to try to make sure it just represents the lasers, not the entire system. The budget is always changing. The federal budget is changing - it's changed again since November. The numbers that we were talking about, the quantities reflect what the budget looked like before Congress and the President got together in November. However, that wasn't a big change, there was about a 5 billion dollar cut out of about $296 billion.

The rest of the presentation is kind of in four blocks. The first block I am going to try to give you an idea of how big the market is ... the military market ... for lasers. The second block I am going to talk a little bit about what the distinction is between a commercial laser and a military laser. And then I am going to get into a little bit more detail of what the characteristics are of the typical lasers that are being used in the military. Then, finally, roll it all together in some trends.

Probably a very useful way to look at the market is to compare it with the commercial market. And in terms of units, there is a major dichotomy here: About three orders of magnitude between how many units the military buys versus how many units are going into the commercial market. But if you look at the number of dollars that go into it, it is about a factor of between 5 and 6. A lot of the numbers that you see published that talk about how much money there is in the military laser market, include a lot of things that you probably aren't interested in. You are interested in production, how much of the money that the government is spending is going into buying production lasers. SDI is interesting to everybody if you read the newspapers, but you are not going to sell SDI all the kinds of lasers that you build. So we took some care to pull out those kinds of numbers.

One way of kind of getting an overview of who in the military uses these lasers is to consider three categories: Strategic Forces, Tactical Forces, and Research and Development. Just about all of the units of lasers go into the tactical forces. Strategic forces are the B52's and the ICBM's and all that sort of thing and they don't use a lot of lasers. In research and development, there is a lot of money, but there are not very many lasers. Most of the lasers that are bought by the military are bought by the tactical forces. I will talk a little bit more about what tactical forces are, but in overview sense it is the guys on the ground, their tanks, the laser designators they carry around in their backpacks, the airplanes that are carrying laser guided missiles ... that is the tactical forces. In terms of dollars, there are more dollars going into research and development, but you are not typically selling high energy lasers. There are only about six programs in lasers in SDIO research that make up most of this money. And they are big demonstrator programs, not buying lasers in quantity. They are demonstrating a great big laser of some kind to see if they can shoot a drone down. There is some research up here, but again, it doesn't represent a lot of lasers.

Finally, in terms of quantifying the marketplace, we looked at what kinds of lasers are sold. He-Ne's make up about 65% of the units that are sold. Solid state lasers are second, around 25% - 30%, but in terms of cost you can see the typical solid state laser is a lot more expensive than the typical He-Ne that the government is buying. When I am saying He-Ne's I am not talking about the lab type He-Ne's, I am talking about the ones that are integrated into systems. In fact, in everything I am talking about, I am talking about from the same point of view that Dr. Ionson was, I am talking about an integrated system. And as I get farther into it, I will identify who I think you want to try to sell your lasers to. You don't typically sell your lasers, from your point of view, directly to the government. You sell them to the guys who integrate the systems into the government. So a little bit later I will show who the dominant players are.

Moving to the second section, I tried to quantify what the marketplace looked like. About 100 million dollars goes into tactical lasers (the kind of lasers that you want to sell), and in terms of money, most of that is solid state lasers. Now I want to kind of distinguish between what a military laser looks like and what a commercial laser tends to look like -- to give you some idea whether you can build a derivative military product from one of your commercial designs.

One of the most important things, even though the military is spending all this money, they don't spend it in one place quite often, at least when they are buying hardware. They typically buy in small lot sizes (Fig. 1). Another important thing is all these sorts of specification requirements tend to drive the cost up. It is normal when you are estimating the cost of a system, which we do some of at BDM, to multiply a commercial price by a factor of 2 - 5 to get the equivalent function for the military. It certainly can go higher, but that is pretty common. The reason that your commercial device that does exactly the same job goes up by a factor of 2 - 5 is because quite often your specs work across a huge temperature range which is very difficult. You may have some high reliability requirements, and from the military's point of view, it is not good enough, or maybe it is not even appropriate to build a very good device. It is appropriate to build a device that you can say on paper is a very good device. From their point of view, higher reliability means that you have done a whole lot of analysis to show what is going to happen over the long term, that you have done some testing to show what is going to happen, and that you tracked your devices once they get into the field. That is a very costly sort of thing that you have to roll into the cost of your device. That gives you an idea of what the difference is between a commercial and a military piece of equipment.

A hand held designator or the laser in a hand held designator, would typically cost between $500 and $3,000, and the total program buy over the lifetime of the program would be somewhere between 30,000 and 60,000 units (Fig. 2). That sounds like a pretty good sized buy, and it actually is. If you look at it on an annualized basis, it may occur over eight to 12 years, so you are really only selling about 3,000 to 7,000 units per year. As you can see, that is on the high end. A lot of the military systems, they buy a lot fewer than that. So that certainly impacts your calculations on what the cost of your unit is if you are concerned about quantity discounts and that sort of thing. You don't have as much quantity to play with when you are dealing with the government.

MILITARY LASERS AND FUNCTIONS

What is the military currently buying? At the high level the military market is divided into two areas: high energy laser weapons and low power lasers. The name doesn't really mean what it sounds like, the distinction is not based on how power, it is based on function. High energy lasers are designed to destroy things with a beam; the low power lasers are designed to be part of a weapon system, where it is not the beam that does the destruction, it is some other part of the system. So this is where you are typically interested, because it is the aerospace kinds of companies that build those high energy lasers. This is where you are going to be working. The two most important areas where the most activity is are in range finders and designators; and guidance systems. There is a man portable system called the MULE that the army uses for "designation". "Man portable"

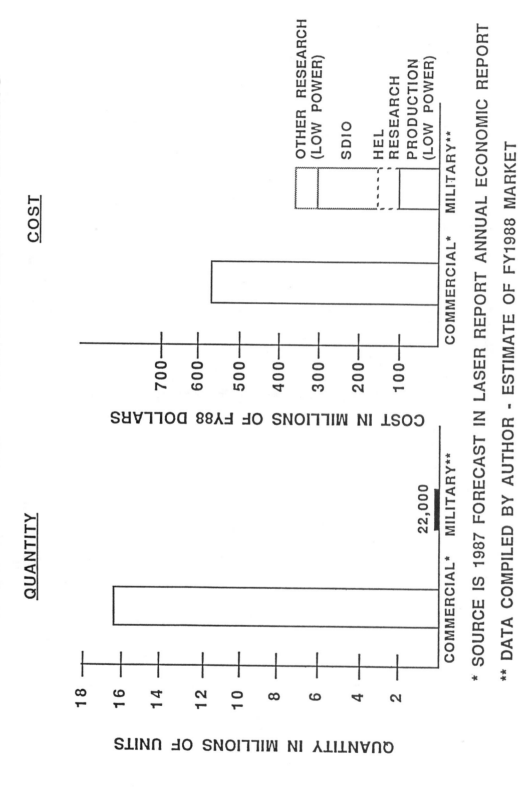

COMPARISON OF COMMERCIAL AND MILITARY MARKETS

QUANTITY

COST

Figure 1

* SOURCE IS 1987 FORECAST IN LASER REPORT ANNUAL ECONOMIC REPORT

** DATA COMPILED BY AUTHOR - ESTIMATE OF FY1988 MARKET

SDIO ESTIMATE EXCLUDES NON-LASER WEAPON OPTICAL EXPERIMENTS

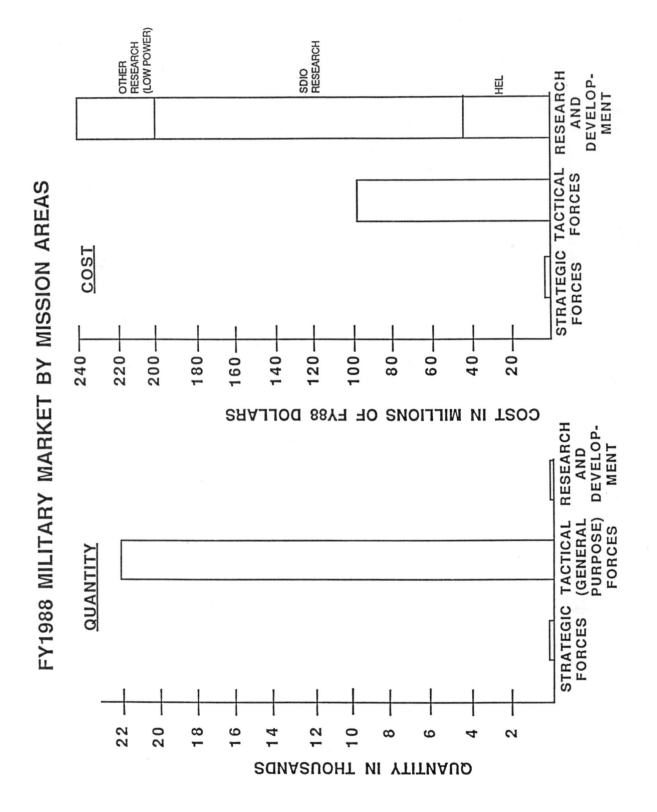

Figure 2

is not a very accurate description of it, it weighs 150 pounds, but
they call it man portable. You go out into the field, and you set it
up, (however long that takes) and point it at a target. It has a laser
in it, and it guides ammunition (like a missile) towards it. So that
is what a designator is. A range finder is exactly what it sounds
like, you aim it at a target and you find out how far away it is.
There is a fair amount of overlap between the functions of these two
systems. A lot of the newer munitions are smart, they have their own
guidance in them. They have their own range finders in them as part of
their guidance, so there is some overlap there. Lasergyros are an
important part of that marketplace as well.

A few parameters important to those systems will give you some
idea whether or not your products fit at all into this marketplace.
Again, the two most important markets are range finders/designators and
guidance systems. Although the range is from 2 millijoules per pulse
to 150 millijoules per pulse, but really most all of the activity is on
the higher end. In both of those systems, the military is interested
in a Q-switch device. They want energy per pulse. They don't need a
real high rep rate, but they do need energy per pulse, and it tends to
be at the higher range. In the guidance system, again, there is some
overlap in these requirements. Low power refers to the gyroscopes. A
lot of the gyroscopes use He-Ne's, so 633nm is also a wavelength that is
of importance to lasergyros.

I have been saying that the guidance systems and the range
finder/designators were an important part of the military market. An
important point to note is that the He-Nes were pulled out of this
quantity because, again, they make up 65% of the market, but a small
part of the dollar, so we wanted to make sure you could see what was
going on down here in the other categories. Range finder/designators
and guidance systems are what lasers are being used for in the military. And they make up about 90% of the total market. In terms of
cost, it is roughly the same, a little bit smaller, about 80% of the
market.

The communication market, I believe, is going to take off. But it
is going to be awhile. The Navy, as Dr. Ionson mentioned before, has
been very interested for a long time in blue/green lasers to talk to
their submarines. The SDIO program is doing a lot of research for all
of these battle stations that they are going to put up in the stratos-
phere to talk to each other, and they may want to use lasers to do that
because they have a whole lot of data that they need to move, so
neodymium YAGs are being considered. A lot of different kinds of
lasers are being looked at, but that is a long ways out. It is 15
years out. So right now the market is not here.

Simulators, that is a steady market. That is where guys go out in
the field and shoot at each other with lasers and they wear vests that
say they have been shot. I have actually been involved with that, and
it is a lot of fun except it certainly excites your neighbors when you
jump out of the back door of your company with an M16 simulator
shooting at each other.

Okay, I mentioned before that typically you don't sell directly to
the government. At least not when you are selling components. The
government may buy components from you directly once the system is
fielded for spares, for initial spares and for replacement spares, but

typically these are the people who sell systems to the government (Fig. 3). If you look at four categories, you will notice that Hughes shows up in every one of those categories. Martin shows up twice. TRW shows up twice. There aren't a lot of players right now in the military marketplace that are selling systems with lasers in them. These are the guys that you want to work with, that you want to sell lasers to, I think. The government, like Dr. Ionson was saying, doesn't buy lasers, they buy the system that does something and it happens to contain a laser. It is very important to keep that in mind.

Figure 3

Major Firms in the Industry 1988

(E/O Products Using Lasers)

HEL RESEARCH
    HUGHES
    LOCKHEED
    TRW
    ROCKWELL ROCKETDYNE
    UNITED TECHNOLOGIES

RANGEFINDERS
    HUGHES
    KOLLSMAN
    ILS

DESIGNATORS
    MARTIN MARIETTA
    HUGHES
    ILS

LASER GUIDANCE
    HUGHES
    ROCKWELL
    MARTIN MARIETTA
    TEXAS INSTRUMENTS

The downside of knowing who these guys are, you can probably also recognize that a lot of these guys make their own lasers. You are going to be competing with in-house capabilities. But if you have a unique system that is much better in some characteristics such as packing density, or power usage or ruggedness, you have the opportunity to compete with in-house capabilities. The Program Manager of Hughes wants to beat Martin-Marietta the next time they bid on a missile. If his in-house capability isn't good enough to get the price down below what he thinks Martin can do it, he is going to look at you if you come in and talk to him. You have a chance, he is not going to be so parochial that he is not going to win the procurement.

TRENDS IN MILITARY LASERS AND MARKETS

We tried to look at this information and other information and put together some thoughts on what the trends are from a qualitative point of view. It is obvious in the literature and in talking to people at the various government labs, that eye-safe lasers are very important to the military. It may seem like an unusual request that a weapon system be eye-safe, but it not really when you think our GI's have to maintain it, our GI's have to be trained with it and that sort of thing. You don't want to be blinding our own GI's. Also, CO_2's, as they are improving they are becoming more popular, especially on tanks. I have seen them on some missile systems because the battlefield conditions of smoke and dust and that sort of thing, 10.6 microns gets through it a lot better than the 1.06nm neodymium YAG, so CO_2's appears to be a growing area. Ring lasergyros are not only replacing the gimbel-type gyros that are already in systems, but are creating new markets. New munitions are being designed around the very compact, easily located ring lasergyros, so anyone that builds a laser that is compatible with ring lasergyros, and there is a wide variety of lasers that are used in those, that is a field we believe is going to go somewhere.

Increased durability is always a trend in the military systems. The military, about 15 or 20 years ago, officially recognized that it makes more sense to spend a little bit more early on in a program than a lot way out in the program to maintain a system. When they select a proposal, or review a proposal on big systems especially, you have to put in a life cycle cost estimate. So if you can show that your system that costs 50% more saves several hundred percent in the long run, they are willing to pay more up front. So durability is always an important trend in any military system, with lasers especially because early ones haven't been all that durable.

From a quantitive point of view, I thought I was going to be unique today in terms of showing a downturn. I guess I am not. A lot of people have been showing downturns. But a downturn in the military sense is different from a downturn in a commercial sense. The reason that this drop of 12%, we think is going to occur between 87 and 88 isn't because the budget is getting cut or anything else, it is because there is only about 4 systems that dominate that number. They have gotten to a point in their procurement cycle that the buy is starting to drop off. No new systems have been designed yet. So this may look like the government is losing interest in lasers, and it is really more of a reflection of understanding the budget process and procurement process. So if you are working in that arena, you need to understand the budget process to some extent, and understand the procurement cycles.

In summary, the observations are again, that marketing the military market takes a different technique than it does marketing the commercial people. You can't create a demand in the military. What you have to do is understand their requirements. Just like Dr. Ionson was saying, you have to understand what that program manager needs. You have to know whether volume is most important to him or power usage is most important to him. What is it that he really needs, and that is what you have to sell him on. Maybe you have to modify your product to meet that need, but it is certainly different from the commercial marketplace where, with the right kind of advertising and that sort of thing, you may be able to create a demand. Cost is not necessarily the primary determinant. There are so many specifications on a military laser, -weight, volume, ruggedness- that sort of thing, that you may be able to compete. You may have a much more expensive laser than a lot of the other people but you may have the only laser that meets all the specifications. So you have the only laser that that integrator can buy when he builds his system. Again, we point out from what we have seen, ring lasergyros are going someplace. If you build a laser compatible with that, you want to talk to the Honeywells of the world and those kind of people who are building those. And, as I mentioned before, it is a limited set of people who integrate lasers, and they do produce their own lasers, so you are going to be competing against in-house capabilities, but it can be done. I don't mean to leave the message that there is no room for newcomers, there certainly is.

AUDIENCE QUESTIONS

Q: I THINK YOU ALSO HAVE A PULL THROUGH EFFECT. I TALKED TO ONE COMPANY ABOUT THE PRODUCT MY COMPANY REPRESENTS, A WRITE-ONCE DIGITAL MAPPING SYSTEMS FOR MILITARY JETS. AND THEY SAID THEY HAVE ALREADY TALKED TO BOEING COMMERCIAL AIRCRAFT COMPANY ADOPTING THAT SAME SYSTEM FOR COMMERCIAL JETS. SO SOME OF THESE PRODUCTS COULD END UP VERY WELL IN THE COMMERCIAL MARKET AS A FOLLOW THROUGH OF BEING ADOPTED BY THE MILITARY.

A: There is always a chance of a derivative product for the commercial market.

JAPAN
THREAT OR OPPORTUNITY?

Joe Nagasawa
President, TEM Co. Ltd., Tokyo
Contributing Editor, Laser Focus Magazine

My topic is Japan, Threat or Opportunity. I hope to present you with some ideas on:
- How big the market is by product areas
- Who the players are in the market
- Who the competitors are (domestic and imports)
- Show some growth trends in various product areas
- The opportunity areas in industrial, scientific, medical, telecommunications, and others
- Where the strength areas of the Japanese manufacturers are
- What we would consider threat areas - the semiconductor lasers, telecommunications, industrial and others.

Then, I hope that you can form your own opinion about whether this is a threat or an opportunity for you.

These figures, (see Figures 1-7), were put out by the OITDA in January 1987. These are basically 1986 figures. Their figures for 1987 have not been released yet. So we will do some 1987 projections. All the numbers are in Yen (I tried to do conversion into dollars, but I don't know what numbers to use), so we will just keep it in Yen and you just use whatever number you want. (The exchange rate has changed dramatically from 250 Yen/$ to 130 Yen/$ over the past several years).

I think you can see that there has been growth. I put up numbers from '83 because OITDA uses these numbers, up to '86, in the general opto-electronics (Fig. 1). We call this production in Japan; these are basically numbers for Japanese manufacturers' production. So the trend is upward. I will try to break this down into three segments.

There are a lot of market reports in Japan and almost all of the reports come out with different numbers. It is very difficult just by looking at reports to see what the real numbers are so we used the OITDA format because they break it down really nicely. Then we tried to qualify those numbers using other reports and by some sampling of the market, talking to manufacturers and users, etc.

Just taking the lasers as a component, the numbers look like this (Fig. 2). This just shows actual production to 1986, so we did a little projection to 1987. There was quite a bit of end-of-the-year procurement forced by the U.S. government pressures, etc. We don't know yet exactly what those numbers were, but most sales were in the R&D sectors. As you can see, there is not too much total growth.

What does the Japanese market look like compared to the U.S. market or the rest of the world? The rest of the world looks pretty much balanced out. The absolute numbers may be disputed, but the percentages are typical. Semiconductor lasers are 28% of the total world market, whereas in Japan it pretty much dominates -- over 80% of the market in semiconductor lasers.

So, who are the players in this market? I have put together a list (Fig. 3) breaking it down by company and by laser type.

Going to growth trends, the growth rate for lasers in Japan, (based on installations) for '85, '86 and '87 (projected) shows there is quite a bit of growth between '86 and '87 for the total market (Fig. 4). Most of it is for the semiconductor lasers. We will go over the breakdown of that in later graphs.

Semiconductor laser production in Japan is presented in millions of Yen); broken down by output wavelength (Fig. 5).

The optical fiber sensor market shows substantial growth, as does the optical disks market (Fig. 6).

For the medical laser systems market, there is about a 21% growth from 1985 to 1986.

For CO_2 lasers (Fig. 7), there was a very minor growth, about 3% (projected) for '86 to '87. Previously '85 to '86 was a pretty good growth year.

The YAG laser market was down quite a bit from '85 to '86. This was considered to be unusual in Japan, so we are not sure what the reason was. But the '87 market does show a little growth. It is about a 3% growth from '86 to '87. I broke that down into different applications, you can see that the semiconductor scribing and trimming market pretty much dominates this market.

The total market for opto-electronics is approaching the neighborhood of a trillion Yen for 1986. That calculates into dollars at over 5 billion, but keep in mind this includes LED's, detectors, CD players, etc., and not just lasers. And there are systems and equipment figures also rolled into this, not just lasers as a component.

AUDIENCE QUESTIONS

Q: WHAT IS THE BIGGEST OPPORTUNITY IN JAPAN?

A: For foreign companies products which are innovative. My perception is the Japanese tend to produce lower cost products once they get started. Foreign products that are innovative have a good market.

Figure 1

Opto-electronics Industry Production 1983-1986

	ITEM DESCRIPTION	1983	1984	1985	1986
OPTICAL COMPONENTS	LIGHT EMISSION DEVICES	118,521	121,826	134,719	159,293
	SEMICONDUCTOR LASERS	15,894	20,550	40,797	54,015
	for communications		15,468	22,739	30,264
	non-comm		5,082	18,058	23,751
	gas lasers	4,999	6,658	8,027	8,887
	solid state lasers	1,282	3,883	3,635	2,935
	LED's	96,271	90,692	82,198	93,386
	for communications	4,202	4,545	5,075	16,034
	non-comm	92,069	86,147	77,123	77,352
	other lasers	75	43	62	70
	LIGHT RECEIVING DEVICES	21,717	21,945	23,946	29,325
	for communication		3,578	5,929	7,233
	individual devices		16,550	12,621	14,480
	arrays		1,817	5,396	7,612
	COMPOSITE OPTO-ELECT	40,190	39,289	32,463	39,186
	SOLAR CELLS	9,068	9,097	10,565	12,002
	OPTICAL FIBERS	39,918	50,525	54,059	67,228
	BeO fibers	37,609	48,265	51,771	64,665
	Non-BeO fibers	2,309	2,260	2,288	2,563
	OTHER OPTICAL COMPOSITES	12,072	36,060	46,371	62,449
	optical connector	2,912	4,667	6,933	8,974
	others	9,160	31,393	39,438	53,475
	SUB-TOTAL	241,486	278,742	302,123	369,483
OPTICAL EQUIP AND SYSTEMS	COMMUNICATIONS EQUIP	21,995	31,859	28,234	36,456
	MEASUREMENT	5,223	16,698	20,717	26,753
	CONSTRUCTION	2,405	4,650	5,024	4,707
	SENSORS FOR FIBERS	1,200	1,368	2,501	4,502
	SENSORS FOR LASERS	4,175	6,994	6,894	8,864
	DISK	46,818	114,817	249,653	300,075
	DAD	28,241	66,994	198,252	236,861
	VD	18,577	43,797	45,783	50,279
	recording disk		3,682	5,130	12,030
	CD-ROM		344	488	905
	OPTICAL IN/OUT SYSTEM	43,109	53,802	75,943	102,440
	MEDICAL	4,567	4,311	4,833	5,829
	INDUSTRIAL	27,355	48,815	52,878	53,570
	SUB-TOTAL	156,847	283,314	446,677	543,196
OPTICAL SYSTEM APS SYSTEM	OPTICAL COMMUNICATION	61,196	73,228	92,662	119,662
	public comm	40,077	39,713	53,154	65,627
	user comm	21,119	33,515	39,508	54,035
	OTHERS	7,176	6,867	6,434	7,226
	SUB-TOTAL	68,372	80,095	99,096	126,888

(ALL VALUES IN 100 MILLION YENS)

	ITEM DESCRIPTION	1983	1984	1985	1986
	TOTAL	466,705	642,151	847,896	1,039,567

SIZE OF LASER MARKET INCLUDING LED's

	1,985	1986	1987 (PROJECTED)
GAS LASERS TOTAL	8,027	8,887	9,190
CO2	3,406	4,261	4,350
HeNe	2,273	2,203	2,280
ARGON	1,636	1,582	1,660
OTHER GAS	712	841	900
SEMICONDUCTOR	40,797	54,015	67,500
LED's	82,198	93,386	104,600
SOLID STATE	3,635	2,935	3,010
OTHERS	62	70	80
TOTAL	134,719	159,293	184,380

LASER PRODUCTION IN JAPAN (MILLION YEN) WITH 1987 PROJECTION.

Figure 2

WHO ARE THE PLAYERS IN THIS MARKET?

LASER SYSTEMS MAKERS IN JAPAN

This list of laser manufacturers was compiled by YANO RESEARCH for the 1987-88 Japan laser report.

LASERS	DOMESTIC MAKERS	IMPORTS/REPS	SYSTEMS MAKER	(APPLICATIONS)
SOLID STATE YAG LASERS	NEC	COHERENT/RIKEI	ABE TRADING COMPANY	(MARKING)
	TOSHIBA	SPECTRA PHYSICS, QUANTARAY, LINE-LIGHT LASER/	TERADYNE	(TRIMMER/MEMORY REPAIR)
	FUJI ELECTRIC	MARUBUN	ESI JAPAN	(TRIMMER/MEMORY REPAIR)
	NIPPON HIGH FREQUENCY	CONTROL LASER/JAPAN LASERS	MARUBENI HI-TECH/CHICAGO	(TRIMMER)
	JAPAX	RAYTHEON/SHINWA TEC	SHINWATEC/BASSEL	(MARKING)
	NIHON WELDING	LASAC/NICHIMEN	NICHIMEN	(CUTTING/WELDING)
	MIYACHI LASER	QUANTRONIX/LEONIX	LEONIX	(MASK REPAIR)
	OSADA ELECTRIC	QANTEL INTERNATIONAL/AUTEX	MARUBUN/QUANTRAD	(MARKING)
	NIIC	SPECTRON LASERS/TOKYO INSTRUMENTS/TEM	SEIBEL	(MEDICAL/SURGICAL)
			M & M	.
			PASCAL SCIENCE	
			MOCHIDA	
			TAKADA/COOPER VISION	(OPHTHALMIC
			CHUO SANGYO	OPERATION
			NIHON LASSAC MEDICAL	DEVICES
			NIHON MDM	MEDICAL
			HOYA	YAG LASERS)
RUBY LASERS	NIHON KAGAKU ENGINEERING	APOLLO LASER		
	NIHON HIGH FREQUENCY			
GLASS LASERS	HOYA	COHERENT GENERAL/CGJ JAPAN		
HeNe GAS LASERS	NEC	MELLES GRIOTEN/NIHON MELLES GRIOTE	YHP	(MEASUREMENT SYSTEM)
	TOSHIBA	HUGHES/KANTUM ELECTRONICS	MECHATROL TRADING	.
	NIHON KAGAKU ENGINEERING	SPECTRA PHYSICS/MARUBUN	SUNTECNO	.
	KIMMON DENKI	COHERENT/RIKEI	SEKI TRADING	.
	USHIO DENKI	UNIPHASE/AUTEX	KANEMATSU ELECTRONICS	(CURRENT METER)
		PMS/JAPAN LASERS	NIHON KANOMATICS	
		SIEMENS/FUJI ELECTRONIC COMPONENTS	LEONIX	.
			KANEMATSU ELECTRONICS	(FLOW INSPECTION)
			SUMISHO ELECTRONICS	(FLATNESS TESTER)
			MATSUSHITA	(CURRENT METER/FA BAR CODE
			KONDO KOGYO	(AIR DUST COUNTER)
			NITTA	.
			ROKOO SHOJI	(MONOCHROME SCANNER)
			EWIG SHOKAI	(FA BAR CODE SCANNER)
			AIDEC CONTROLS	
			MOCHIDA PHARMACEUTICAL	
			TAKADA COOPER VISION	
			CHUO SANGYO	
			(PHOTOCOAGULATORS)	

Figure 3 (page 1 of 3)

	LASERS	DOMESTIC MAKERS	IMPORTS/REPS	SYSTEMS MAKER	(APPLICATIONS)
	ARGON	NEC	COHERENT/RIKEI	OSAKA YAMATOYA SHOKAI	(COLOR SCANNER)
	(WATER	TOSHIBA	SPECTRA PHYSICS/MARUBUN	KAIGAI SHOKAI	•
	COOLED)		LEXEL/LEONIX	NIHON SAIKEKUSU	(PHOTO PLOTTER)
			AMERICAN LASER/KANTUM ELECTRONICS		
			CONTINENTAL LASERS/AUTEX		
			SPECTRA PHYSICS/SPKK		
	ARGON	TOSHIBA	SPECTRA PHYSICS/MARUBUN		
	(AIR	NEC	OMNICHROME/NIHON MELLES GRIOTE		
G	COOLED)		AMERICAN LASERS/KANTUM ELECTRONICS		
A			CATHODION/RIKEI		
S			ION TECHNOLOGY/JAPAN LASER		
			SIONICS, CONTINENTAL LASER/AUTEX		
	KRYPTON		COHERENT/RIKEI		
			SPECTRA PHYSICS/MARUBUN		
			LEXEL/LEONIX		
L	HeCd	KIMMON DENKI	OMNICHROME/NIHON MELLES GRIOTE		
A			LICONIX/SEKI SHOJI		
S	CO2	MITSUBISHI	PHOTON SOURCES/SHINWATECH	AMADA	
E		TOSHIBA	COHERENT GENERAL/CGI JAPAN	ANRITSU	
R		MATSUSHITA	SPECTRA PHYSICS/SPKK	DIHEN	
S		NEC	ROFIN SINAR/MARUBENI	HITACHI	
		HITACHI SEISAKUSHO	MESSER GRISHEIM/KOIKE SANSO	IWATANI SANGYO	
		DIHEN	TRUMPF/TRUMPF JAPAN	JAPAX	
		SHIMADA RIKA	PRC/MIYAMA/SHIBUYA	KOIKE SANSO	
		NIIC	FERRANTI/NICHIMEN	KOMATSU	
		FANAC	AVCO, APOLLO/JAPAN LASER	MARUBENI/ROFIN SINAR	
		NIHON KAGAKU ENGINEERING	LUMONICS/LEONIX	MATSUSHITA	
		KAWASAKI HEAVY INDUSTRY	LUMONICS/SHINWATEC	MITSUBISHI	
		MOCHIDA PHARMACEUTICAL	LASER APPLICATIONS-COHERENT/EDM	MIYAMA	
		ALOKA	CE/?	MURATA	
		YOSHIDA SEISAKUSHO		NIHON INFRARED INDUSTRIES	
				NIHON MESSER GRIESHEIM	
				NIHON SANSO	
				NIIGATA TEKKO	
				NIPPEI TOYAMA/ROFIN SINAR	
				NISHIN BOEKI	
				SHIBUYA	
				SHIMADA RIKA	
				SHIN MEIWA	
				TOSHIBA	
				TRUMPF	
				URAWA	

Figure 3 (page 2 of 3)

LASERS	DOMESTIC MAKERS	IMPORTS/REPS	SYSTEMS MAKER	(APPLICATIONS)
GAS CO2 (CONT)			YAMAZAKI/MAZAK	
			MARUBUN (MARKING)	
			C. ITOH (MARKING)	
			SHIBUYA (MARKING)	
LASERS TEA CO2	SHIBUYA KOGYO	LUMONICS/LEONIX	MITSUI BUSAN/XMR	
	NEC	LUMONIX/SHINWATEC		
	TOSHIBA	LASER APPLICATION-COHERENT/E.D.M.		
	USHIO ELECTRIC			
	KYUSHYU ELECTRONICS SYSTEM			
	FUJI ELECTRIC			
	IZUMI ELECTRIC			
EXCIMER	HAMAMATSU PHOTONICS	LAMBDA PHYSICS/MARUBUN		
	SHIBUYA KOGYO	LUMONICS/LEONIX		
	KOMATSU	QUESTEC/SPKK		
	NEC			
	NIHON DENSHI			
SEMICONDUCTOR VISIBLE WAVELENGTH	SHARP			
	MITSUBISHI			
	MATSUSHITA			
	SONY			
	NEC			
	HITACHI			
	TOSHIBA			
	ROHM			
	MITSUI			
	SEIKO EPSON			
	TATEISHI			
LIQUID DYE		COOPER LASER SONICS (LEONIX) •		
		QUANTRONIX		
		LUMONICS		
		CANDELLA		
		SPECTRA PHYSICS (MARUBUN) •		
		QUANTA RAY		
		LAMBDA PHYSICS		
		COHERENT/RIKEI •		

Figure 3 (page 3 of 3)

Figure 4. Growth rate for Lasers in Japan.

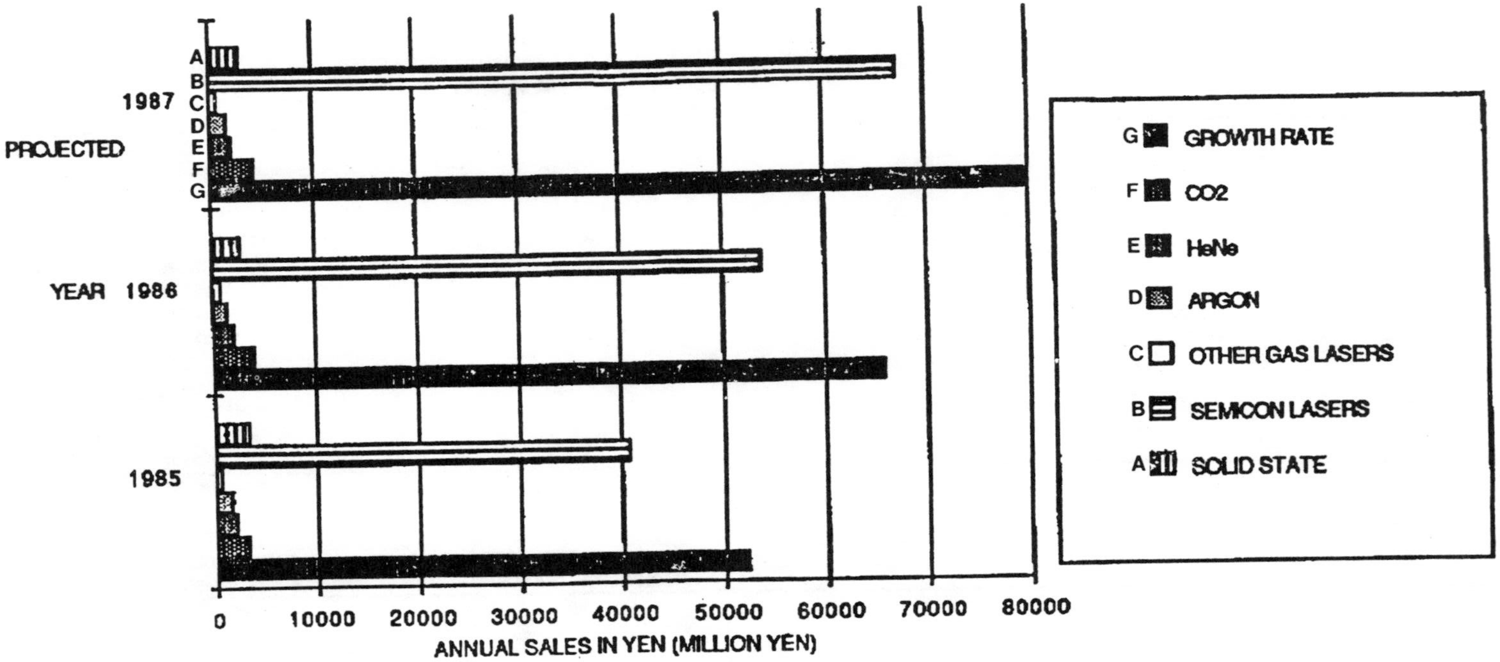

GROWTH TRENDS BY PRODUCT AREAS

"Growth rate for lasers in Japan". This graph shows the annual sales figures in YEN of the laser devices installed in the Japanese market. The total growth rate from 1985-86 was around 20%. Semiconductor lasers accounts for the main portion of this growth of this market as well as the major market share.

Figure 4

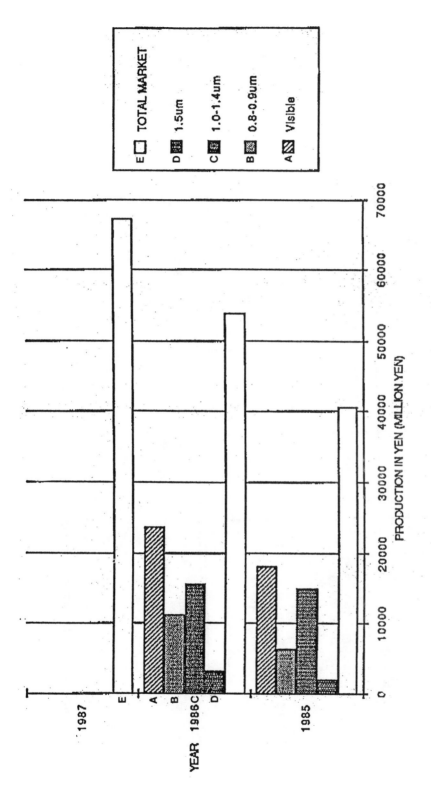

"Semiconductor Laser Production in Japan". This graph shows the breakdown by wavelength of the semiconductor laser market in Japan in YEN. In 1986 the 1.3um contributed about 16 billion YEN to the total production and in the future the 1.5um laser for long distance communication and the 0.8um laser for local loop communication will contribute to the growth of this market. The total market growth in this field is estimated to be between 25-30% annually.

Figure 5

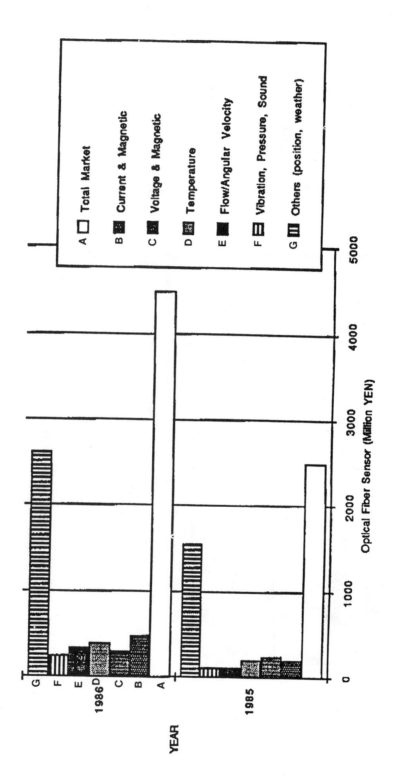

Figure 6

"Optical Fiber Sensor". This graph shows where the fastest growth was in this market, where the growth was around 80-83% annually.

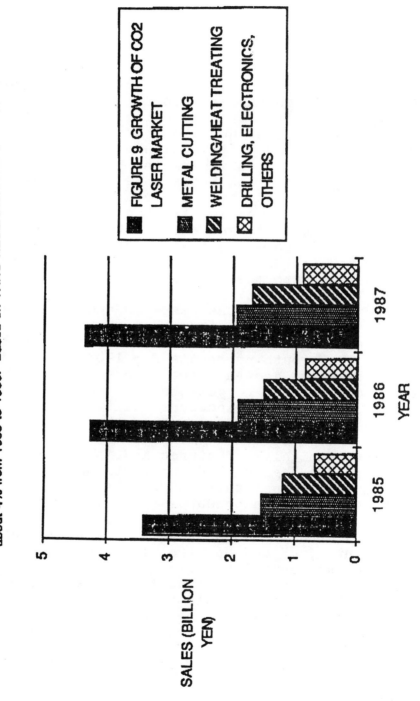

"GROWTH OF CO2 LASER MARKET". This chart shows the actual growth from 1984 through 1986. The growth for CO2 was only about 4% from 1985 to 1986. Based on YANO RESEARCH REPORT 1987-88.

Figure 7

EUROPE
AN INSIDER'S VIEW

Paul Crosby
Sales Director, Laser Products Division
Coherent

What I would like to do is to really answer a question which most American companies find themselves wrestling with when they first start to consider the European market. That question is, "should one view Europe as a single entity, or as a collection of individual states?" Once you have answered that question, then from that is driven your whole marketing sales and distribution policy.

What I would like to do is examine three of the major segments of the laser market, that is the research & development segment, the medical laser segment, and the industrial laser segment.

I will also spend some time talking about Eureka, which is the European answer to SDI. As you know, both Japan and Europe viewed the funding that the U.S. government was putting into SDI as threatening, long-term, the manufacturing industries because of spin-off technologies. So I will go into the Eureka and your laser program, and what impact that will have on it.

My figure (Fig. 1) for the research and development market in Europe is a little bit higher than that that Jon Tomkins had. I estimated it as being $65 million, and constituting 35% of the total world market. The growth prospects for 1988 are good, running around 10%. This ignores any growth associated with the stronger European currencies. Sometimes we forget that the markets actually grow as the currency diminishes. The price sensitivity for most high capital cost equipment is set in local currency, and there is not the same expectation for those prices to track with the diminishing dollar.

The market is highly structured and is dominated by Lumonics and JK, Quantel, Spectra-Physics and Coherent/Lambda Physik. Together these four companies constitute 75% of the total market. I estimate that the U.S. manufactured products for this contribute 65%. Those areas that are being manufactured in Europe are principally in the post laser technology areas. Lambda Physik in Germany manufacturing excimer dye lasers, Quantel, JK and Spectron in France and U.K., with Oxford lasers making metal vapor lasers.

If you examine the relative dollar market values of each of the European states, what you find is that Germany is by far the largest country. This somewhat belies the truth of the matter, and the amount in what one would consider true research and development in both universities and in the industrial area is larger still. Germany doesn't have the same military or aerospace complex that both France and the U.K. have, so the true size of that research and development community as such is significantly larger than 30%. Italy is growing substantially. It now has a Gross National Product which is higher than the U.K., although the purchasing patterns in Italy are sporadic at best. One gets business about every three years or so. But when it comes, it is really worth having. The markets are not totally homo-

Figure 1

The European R&D Market

RELATIVE $ MARKET VALUE

% OF EUROPEAN MARKET

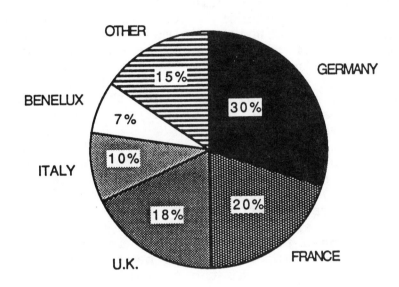

geneous. There is some loyalty to nationally produced products, I am sure most people that are in the market will be aware the laser market in Germany is dominated totally by excimer lasers, and the market in France is dominated totally by YAG lasers. These historical trends are slowly disappearing. One is seeing more homogenization, but certainly in France and Germany the situation is still polarized to the extent that one would anticipate.

Some of the interesting trends that are occurring in Europe: the creation of international research projects sponsored by the European Economic Community. The ones that immediately spring to mind are BRITE and Espirit. BRITE being geared towards doing basic research for industrial technology and Espirit being an information technology initiative. These projects are making it necessary for countries to operate in joint research areas so that Spain must cooperate with France, and Italy must cooperate with Germany in order to get funding under them. So there are initiatives being taken to break down the cultural barriers within Europe.

Other things that will happen is, the countries that we normally consider to be outside the mainstream of the business, Spain, Portugal, Greece, will become more significant over the next four to five years, as EEC funds are administered more equally.

My final conclusion is that the U.S. manufacturers continue to dominate the R&D market, because the American scientific market is strong enough to support new developments. The R&D market is terribly innovation driven. If you have a new product then you can go and sell it in the rest of the world. But if you are trying to make new products in Britain or in Japan, the size of your own domestic market is finite, and making that next jump is very, very difficult indeed. So, suppliers to the R&D community, not only in Europe but worldwide, -- American manufacturers are relatively secure with the proviso that we actually manage the business. It is my contention that as suppliers to both Europe and Japan, which the most conservative of estimates constitutes over 50% of our business, we expend something approaching 10% of our time on. And certainly the time that is spent, is spent within the marketing and sales group and not in the whole company. If we wish to maintain this dominance, then the whole company must think of the global R&D market and not think parochially in such terms of only 'domestic' market needs.

I would like to talk a little bit about BRITE and Espirit. The BRITE project areas are in reliability of 10 kilowatt lasers, new welding processes, new testing procedures (with an emphasis on non-destructive testing). In order to qualify for BRITE funds, the project must be of four years in nature, 50% of the project must be paid by a sponsoring industry, and there must be two countries involved. So far, 100 projects have been funded, each to a level of 1.5 to 2.5 million ECU's. So far the amount of money we have spent is 100 million.

Now, if one is to look at the scientific market, or the R&D markets in terms of whether you should treat Europe as a group of state or as a homogeneous community, the conclusion that one comes to is that as a market, it is an homogeneous market. There are two caveats to that - and that is you must distribute differently in each country and be aware of the cultural differences, but as a market in itself it is a single entity.

THE MEDICAL /OPHTHALMIC MARKET

I would like to compare and contrast that with the European ophthalmic market, which has an overall size of 20 to 22 million dollars. We expect very little or no growth in this traditional area. The only growth anticipated in photo-coagulation in Europe is in endo-photocoagulation, which is in procedures relating to the front of the eye. There is a ratio of photo-coagulations as opposed to disruption, or argon to YAG treatment of approximately 17 to 30.

If one examines the relative dollar market values, it is immediately apparent that France is the biggest country (Fig. 2). Italy is a lot larger than one would anticipate just on the basis of population. This is because of the different patterns of medicine that go on in these two countries compared to the other European states. France and Italy have an office photo-coagulator market, that is the ophthalmic surgeons who can use laser surgery, can treat people as out-patients and can be directly reimbursed. Whereas the U.K. and Germany have

Figure 2

Medical Ophthalmic

RELATIVE $ MARKET VALUES

RELATIVE SIZES OF OPHTHALMIC MARKET

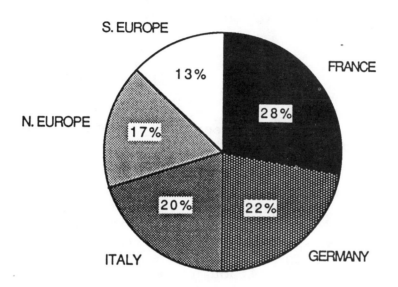

strong socialized ophthalmic programs. You can imagine that if there is a program of social ophthalmic medicine in Germany, then it would run moderately effectively.

European manufacturers totally dominate the photo-coagulator business. Meditec Science and Rodenstock producing both YAG and argon photo-coagulators. LaserTek producing both. And Bio Physik, Cilas in France also, and Lasay in Switzerland. The dropping of the dollar does make it more attractive for U.S. manufacturers as a marketplace, but the prices in Europe have been lower on average than the prices recognized in the U.S. market.

One of the barriers for a U.S. company in operating in these medical areas is the requirement for the wiring, or the safety shutter-ing or the labeling to be totally in compliance with the local legisla-tion. In the photo-coagulator area, the European community has worked to a standardization. So the threshold in getting into the photo-coagulator area is not as great as it is in the surgical area, which I will come to later on. But it does mean, if you are exporting goods, you do have to build them and wire them differently, and color code them differently.

The U.S. does participates in the argon photo-coagulator business, being the only source of the Argon lasers that are used in the photo-coagulators in Europe. (Or I should say greater than 90%).

The trends in Europe that we anticipate are, first the photo-coagulator market will be primarily a replacement market and there have been some new entrants in Europe over the past year with both Zeiss and Rodenstock entering into the market. This is rather difficult to come to terms with when one views what has been happening in the U.S. market, but both Zeiss and Rodenstock, I believe, view the lasers as accessories to their main product line, and one can rationalize their decision from that point of view. It is difficult to see how Europe can sustain as many indigenous laser manufacturers profitably, given the size and the strength of the two new entrants.

THE SURGICAL MARKET

Moving on to the medical/surgical area, which has a total market of 15 to 18 million, there is a very large growth potential of up to 25% in certain segments, with 15% overall. There are two major technologies in medical lasers, they are CW YAG or CO_2 lasers. Broadly speaking, CW YAG is used as a deep heater, and is used in such thera-pies as bleeding ulcers or some areas in heating and removing parts of tissue. Whereas the CO_2 laser is essentially operating as a laser scalpel. The greatest growth is expected in CO_2.

When one looks at the medical/surgical market, it is characterized in Europe by tremendous amounts of regional variation - from state to state. If you stop and consider the critical factors that affect the acceptance of a laser procedure in each country, these are the basic histories of medicine in the country. At the moment there is no European safety standard, although one is being worked on for lasers to be incorporated into operating theaters. You also have new surgical procedures that the clinical community is favoring. The clinical community is very small, and is almost xenophobic, in that it only talks to people within its own area. So the German clinical community only talks to other people within the German clinical community.

And finally, you have to look at the different reimbursement patterns as compared to the U.S.. I can perhaps illustrate this best by looking at how you would treat a cervical inter-epithelial neoplasia in three different countries. In France one would go for cryosurgery, that is the accepted way for carrying out a procedure of that nature. In Germany it would be CW YAG. In the U.K. it would be CO_2. In the U.S. it would be a CO_2 procedure or a straightforward surgical pro-cedure. So you can see that in coming to terms with this, one must look at the market country by country in the surgical area.

If you look at the status of the two different treatments, in Germany it is mature CW YAG markets and the emerging CO_2; France is growing in both and has been possibly the slowest of the European countries to come to terms with laser surgery; Italy is growing in both areas; and the U.K. is relatively mature in both, but is also is of rather small volume because of the impoverishments in the National Health Service. It is no great surprise that when you actually match up the procedures that the manufacturers fall into the same areas, and that is that the principal manufacturer of CW YAG in Europe is MBB, and we are seeing some growth of CO_2 manufacturers in Zeiss, Cilas, and

Laser Applications Ltd., which was acquired by Coherent this year. The primary supplier to the growing CO_2 market has been Laser Industries, or Sharplan.

On the more esoteric areas, Germany this year has founded three new centers for lasers in medicine, Berlin, Lubek and Ukm. These are to take top quality surgeons, top quality physicists and to really to try to develop some new surgical procedures. We believe that some of the areas that they will be looking at are laser lithotripsy, which is the kidney stone smasher; corneal sculpturing; laser angioplasty; and also photo-radiation therapy, but photo-radiation therapy has largely been tackled as a research area that Italy has chosen as its own.

If you now look at how you should view the medical market, it is much more segmented in terms of its states, and you really have to look at it as being a whole group of separate markets within Europe. In surgical in particular, there are tremendous barriers for a U.S. company in entering that area because of the non-unification of the safety standards in operating hospitals, although this is coming.

THE INDUSTRIAL MARKET

I would like very briefly to review the European industrial market. These figures are taken directly from the recent Laser Focus article. The European market is estimated as being 25% of the world market, with a growth rate of 13%. It seems that the market will continue to develop, especially in the CO_2 area. It is significant here that with the exceptions of both Lumonics and Spectra-Physics, the vast majority of the European market is supplied by European manufacturers, or manufacturing subsidiaries of U.S. companies.

If one looks at the manufacturers, then Rofin-Sinar and Trumpf together are estimated to cover 60% of the European market; with Coherent General, Cilas, Ferranti, and some manufacturers in Italy making up the balance. This year, the acquisition of Rofin-Sinar by Siemens will probably have the greatest single impact on the CO_2 laser industry worldwide. Siemens is spending a great deal of time and aligning themselves with laser technology. Furthermore, the acquisition of Sciacky by Ferranti will make European manufacturers a more formidable contestor in the U.S.. Within Europe at the moment, there is strong funding for the development of high power industrial lasers. These are government funds which are being directed to produce CO_2 lasers in the 10 kilowatt to 200 kilowatt regime, and excimer lasers which are in the multi-kilowatt regime.

The bulk of this funding comes under the Eureka project. The Eureka project has been put together by the EEC to stimulate innovative small and medium sized companies to make it essential for small companies across country boundaries to work together. The fund is administered by a central coordination office in Brussels. There are four key areas that they are working in. Laser technology and applications; CADCAM software; high definition TV; and broad band telecommunications. The rules are that at least two countries have to be involved, so if I am a manufacturer in Britain then I must work in conjunction with a manufacturer in Italy to submit a proposal towards Eureka and then if that proposal is accepted, then I will matching funds from my government, and the Italian group will get matching funds from their government, to sponsor the project.

So let's take a brief look at the Eureka laser related projects (Fig. 3). The Eurolaser project, which is being primarily driven by West Germany and France, is geared towards high-powered CO_2 and excimer lasers, material processing and information systems. There is a program for producing 25 kilowatt CO_2 lasers cells, when they are stacked together will make it possible to produce a 200 kilowatt CO_2 laser with good beam quality. There is also a program going on in Belgium, Austria and Spain to get improvements in optical efficiency to bring down both the size and increase the efficiency of a CO_2 laser with the output power in the order of 10 kilowatts.

In the excimer area, there is a program which is to develop an industrial test station for excimer lasers, and involved in this project are Lambda Physik, KDU (a subsidiary of Siemens), Sopra, and Oxford Lasers. The other objective behind this program is to develop an infrastructure throughout the industry that can support an industrial excimer base. There is also a program in France to develop high power solid state lasers.

Producing these wonderful high power solid state lasers and high power CO_2 lasers is a very laudable and interesting goal, but if you talk to most of the people who are involved in the CO_2 laser area, they say well, if you have 25 or 50 kilowatts of power, exactly what do you do with it? At the moment most of the unit sales seem to be in the 1 to 3 kilowatt regime, and very little in the 5 to 10 kilowatt. Some of the rationales for that are because of the associated deterioration in beam quality that one gets when you go to a larger CO_2 laser. And some of it is also the applications have just not been developed. So, the second wing of the Eurolaser Project is actually having made these lasers to look for things to do with them.

The projects that are currently funded are: With Belgium, France and Italy to look for lasers in the disposal of toxic chemicals (using high power lasers). There is an application roundtable which is mainly based on information exchange. And an area on beam guidance and control systems.

In terms of assessing the impact of the Eureka program, it is very difficult to do that because if one looks at the historical collaborative projects that have gone on with Europe, you can point to things as to being technologically wonderful, but commercial failures, such as Concord. Another one such as the Airbus that has worked reasonably well, and CERN that has worked reasonably well. It is too early to say whether the laser projects which are geared totally at sort of refitting European industry with the next generation of lasers for manufacturing will work or not.

EUREKA LASER RELATED PROJECTS

EUROLASER	W. GERMANY FRANCE	$95M	HIGH POWER CO_2 & EXCIMER LASERS, MAT'L PROCESSING & INFO SYSTEMS
25 KW CO_2 LASER	U.K., SPAIN DENMARK	$8M	GOOD BEAM QUALITY: 25KW MODULES TO PRODUCE UP TO 200 KW.
INDUSTRIAL CO LASER	U.K., NETHER-LANDS, SWEDEN	$0.6M	CO LASER REPL. CO_2 OR ND:YAG
10 KW CO_2 LASER	BELGIUM, AUSTRIA, SPAIN	$32M	IMPVMNTS IN OPTICAL EFFIC-IENCY, NEW WELDING HEAT TREATMENT.
HIGH POWER EXCIMER	W. GERMANY FRANCE, U.K. NDL	$16.8M	DEVELOPMENT TEST STATION INDUSTRIAL EXCIMERS. L-P KWU, CILAS, OXFORD LASERS DEVELOPMENT ON INDUSTRY INFRA-STRUCTURE
HIGH POWER SOLID STATE	FRANCE, W. GERMANY	$21.7M	MULTI KW ND: YAG LASER W/ GOOD BEAM QUAL. HAAS-LASER, QUANTEL MAJOR CONTRACTORS
LASERS FOR DISPOSAL TOXIC CHEMICALS	BELGIUM, FRANCE NDL, ITALY	$10.3	USING HIGH POWER LASERS
LASER APPL'N ROUND TABLE	W. GERMANY SWITZERLAND	$8.5M	INFORMATION EXCHANGE & FEASIBILITY CENTRE
INDUS. APPL'N HIGH POWER LASERS	BELGIUM	$43M	BEAM STEERING GUIDANCE AND CONTROL SYSTEM
LASER WORK STATION SURFACE TREAT-MENT	ITALY, W. GERMANY	$20.5M	OPERATION OF UNIFORM BEAM PROFILES FOR HEAT TREATMENT RUFIN-SINAR, PRIMA-INDUSTRIES

Figure 3

AUDIENCE QUESTIONS

Q: THE COMMENT ON THE ADDITIONAL FUNDING FOR "PRODUCT AND PROCESS
      DEVELOPMENT" THAT GOES ON, FOR EXAMPLE STATES IN GERMANY WHO HAVE
      INDIVIDUAL PROGRAMS TO SUPPORT INDUSTRY IN THEIR OWN STATES, OR
      THE CONTINUING EFFORTS TO BUILD ACTIVITY IN OTHER COUNTRIES, THE
      ONGOING PROGRAMS IN THE U.K. FOR EXAMPLE THAT ARE SEPARATE FROM
      EUREKA AND BRITE. AM I CORRECT IN SAYING THAT THOSE NUMBERS
      APPROACH 2.5% OF THE EUREKA AND BRITE NUMBERS.

A: They are not included in the Eureka and BRITE numbers. I don't
have a handle on the value of those.

Q: I DON'T SEE HOW YOU CAN COMPARE EUREKA TO SDI. THEY ARE TOTALLY
      DIFFERENT.

A: I don't compare Eureka to SDI, but the EEC. They compare Eureka
to SDI. They see SDI as being a major influx of government expenditure
in the U.S. technological companies which will give rise to industrial
spin-offs. That then means that if they want to keep pace, and they
are under pressure to keep pace, they must fund to a certain extent,
rather than invest the money in the military. That would be a very
difficult and torturous thing to do with it in Europe. They decided to
invest it in manufacturing processes.

Q: WHAT SHOULD A U.S. COMPANY DO TO TAKE ADVANTAGE OF THE RELATIVE
      RISE IN THE VALUE OF THE EUROPEAN CURRENCY TO THE DOLLAR?

A: I think what one has to do is to look at your prices in the
marketplace in local currency. Most companies of a small size tend to
sell on a certain percentage below U.S. domestic. That price is not
adjusted relative to the value of the dollar in the different
countries. If you are using a rep network, you normally expect the
price to ride as a with the dollar. I think you have to give closer
management to your representative network and become more involved in
pricing with them on a day to day basis.

Q: DO YOU THINK YOU SHOULD TAKE ADVANTAGE OF THE GREATER PROFITS THAT
      COMES FROM GETTING MORE DOLLARS FOR YOUR PRODUCT, OR PRICE LOWER
      TO GAIN MARKET SHARE?

A: It is a very difficult to call. You must really decide what your
market position is as to whether you wish to use this as an opportunity
to gain higher market share or whether you wish to gain profitability.
I think my point is that very often we tend to take the decision by
default rather than to investigate it and then decide what to do.

CLOSING REMARKS

Dr. Morris Levitt
Editor-In-Chief, Laser Focus Magazine

We had two broad purposes today. They both serve us as marketeers and technologists. One is more purely intellectual; we all need to know the terrain on which we operate. Given the fact we all have hectic schedules through the year, I see this kind of meeting as an opportunity where we can all take a day and step back and give ourselves an overview of the industry and markets in which we work. There may not be any immediate, practical outcome to that, but I think it's impossible to function effectively without a coherent overview of the field in which we work. I hope the combination of talks today have proven that despite the complexity of the field, it is possible to present a reasonably valid snapshot of the field. In that sense it is appropriate we are here in the context of the SPIE O-E/Laser meeting. You can think of this as an extension course of the type you might take closer to home in an MBA program for technology managers. Therefore, I am proud to award each and every one of you the degree of Master of Laser Marketing.

The second purpose we have is much more directly related to your business and marketing activities. If you were coming here looking for specific information, I hope we either gave you the answers or at least an idea of how to pursue the answers. I think from a forum like this it is possible even in hearing talks on fields that may be distant from your immediate concerns, to be stimulated by new ideas that feed back into your more immediate area of concern. Discussion among all of us creates a level of dialogue which enriches all of our activities.

Finally there is the social component, which is to say, the opportunity for networking, the sharing of various resources we all bring to this meeting. In that sense, I hope we can break down further any barriers between listeners and our speakers. It's critical to have time not only for questions, but for strategic and tactical discussions.

This is the first time this seminar has been presented completely under the aegis of Laser Focus. We are very grateful for your comments. I can only promise you we will make every effort at future seminars to present them in the most professional and accessible way possible.

In addition, we have two more seminars this year. In April, at the American Society for Laser Medicine meeting, we will spend a day under the chairmanship of Mike Moretti looking at all facets of the medical laser market. In September, at the Chicago Medicine Tool Show, we will be with Dave Belforte at the point of production with a full day on the Industrial Laser marketplace.

AUTHOR INDEX